URBAN ENVIRONMENT MANAGEMENT

LOCAL GOVERNMENT AND COMMUNITY ACTION

ISS Publications under Urban Studies Series

1. Urbanisation in Developing Countries: Basic Services and Community Participation
2. Urbanisation in India: Basic Services and People's Participation
3. Basic Services for Urban Poor: Case Study of Baroda, Bhilwara, Sambalpur and Siliguri
4. Urban Poverty, Basic Services and Environment
5. Plague in Surat: Crisis in Urban Governance
6. Mayor-in-Council System in a Comparative Perspective

Archana Ghosh is the Head of Urban Studies at the Institute of Social Sciences, New Delhi. She has to her credit a number of study reports and research articles on decentralization and urban governance, basic services, environment and urban poverty. She has participated in several national and international conferences on these issues.

Mrs. Ghosh has co-authored two books: *Basic Services for Urban Poor: Case of Baroda, Bhilwara, Sambalpur and Siliguri* and *Plague in Surat: Crisis in Urban Governance*

The Institute of Social Sciences, the result of an intellectual quest to provide socially relevant research, aims to study contemporary social, political and economic issues, with an inter-disciplinary perspective. The Institute makes its research findings and recommendations available to government bodies, policy makers, social scientists and workers' organizations so as to widen their options for action. It monitors public policy, takes up relevant research and engages in consistent advocacy. The Institute has since its inception in 1985 sought to bring to the centrestage the issue of participatory local government, people centric development, women's political empowerment and democratic governance. It has a strong International Studies department building partnership with international organisations, institutions and research centres having common interests. South Asia is another major study area.

The Friedrich-Ebert-Stiftung was established in 1925. As the oldest political foundation in Germany it is a private public-interest institution committed to the ideas and values of social democracy. It carries the name of Germany's first democratically elected president, Friedrich Ebert, and continues to pursue his legacy of building freedom, solidarity and social justice. The Foundation carries out this mission, both in Germany and abroad, through its programmes of civic education, international cooperation, scholarships and research.

Urban Studies Series No. 7

Urban Environment Management

Local Government and Community Action

Edited by

Archana Ghosh

Published for

Institute of Social Sciences

&

FRIEDRICH EBERT STIFTUNG

By

CONCEPT PUBLISHING COMPANY, NEW DELHI 110059

ISBN 81-8069-040-7

First Published 2003

8 Nelson Mandela Road
New Delhi 110 070 (India)

Tel: 91-11-26121902, 26895370
Fax: 91-11-26137027
Email: iss@nda.vsnl.net.in
www.issin.org

Printed and published by:

Ashok Kumar Mittal
Concept Publishing Company
A-15/16, Commercial Block
Mohan Garden
New Delhi 110 059 (India)

Tel: 91-11-25351460, 25351794
Fax: 91-11-25357103
Email: publishing@conceptpub.com

Foreword

Growing urbanisation and industrialisation in developing countries has tended to go hand in hand with a deterioration in the environmental status of many of the urban centres. The urgency for managing the degradation of the environment and reducing its impact on public health, productivity and the quality of life has been expressed at several world fora in recent times. The development of key sectors of growth, such as transport, tourism, energy and industry, has seriously affected the quality of the environment at the urban local level. Therefore, it is at this level that implementation of environment policies on air quality management, waste disposal, waste water treatment, etc., should begin. The municipalities being the institutions of self-governance closest to the people have to be prepared to meet the environmental challenges that cities and towns pose today. Civil society organisations, especially NGOs and the urban community as a whole, also have a crucial role to play in protecting the local environment.

These issues have been comprehensively dealt with in the current volume by authors from different countries. *Urban Environment Management: Local Government and Community Action,* the third volume in the Urban Studies Series of the Institute, is the outcome of an international conference held in Kolkata (previously Calcutta) in November 2001 and jointly organized by Institute of Social Sciences and Friedrich Ebert Stiftung (FES). Kolkata has a special significnce as the venue of this conference. Besides being a historic city of great diversity, it is the nodal center of the eastern region of India. Especially impressive is the fact that during the last three decades, Kolkata has been the theatre of massive metro development planning, with a special focus on slum improvement. The city has always been a ground for new experiments in development planning. Lately, democratic decentralisation has become the major thrust area of urban governance. The Mayor-in-Council system, or the cabinet form of government, in the Municipal Corporation of Kolkata, implemented by the city since 1984, has become a role model for other municipal bodies in India today.

The international conference had the active participation of experts from several countries. It was pioneering in the sense that for the first time, mayors of several cities, and specialists from, Bangladesh, Pakistan, Nepal, Sri Lanka, Thailand and Germany, besides India, shared ideas and exchanged views at a common forum. Political leaders and policy-makers from West Bengal, grappling with the gigantic problems of urban governance and attempting to find solutions, enthusiastically took part in the deliberations.

I am thankful to Friedrich Ebert Stiftung for its active collaboration in organising this conference. The Institute has had a long-standing association with FES in the field of urban studies. It has been responding positively to new ideas on conferences, studies and other programmes initiated by the Institute. Several publications and papers are the outcome of our partnership. The present volume is another valuable addition to our collaborative efforts. Mr Manfred Haack, the present Representative of FES in India, has from the very beginning, taken a special interest in generating ideas and planning the logistics and design of the conference.

Some of the successful case studies, included in the book, on local government and community initiatives in environmental management will stand out as examples worthy of emulation by urban governments and civil societies around the world. I am sure this volume, painstakingly put together by Mrs Archana Ghosh, Head of the Department of Urban Studies of ISS, will help to disseminate the new ideas discussed at the conference for providing solutions to the environmental problems plaguing the cities of the world today.

George Mathew
Director
Institute of Social Sciences

New Delhi
1 January 2003

Preface

The Globalised City

In a way, hospitality is one of the most salient features of cities and has been so in the past. During feudal times in Europe, for example, a common saying was "City air makes one free." This meant that particularly in rural areas, where the masses lived as serfs deprived of human rights, those who could escape and finally found a place in one of the bigger cities were granted citizenship and subsequently got new options in life. At that time, imperial free cities with their premature republican constitutions were areas different from the commonly autocratic entities in Europe.

The rural exodus, which started in medieval times, continues up to the present day. It has left village settlements with a somewhat provincial atmosphere and given cities, from the beginning, a more cosmopolitan character. World-wide, urban populations naturally comprise a variety of ethnic and religious groups forming a complex social structure. In this respect, and despite a certain degree of ghettoisation, a modern city is a melting pot of cultures.

Some cities which were founded a long time ago achieved political independence in the ancient world mainly by their economic might. They exploited their geographic advantages, particularly as seaports, and became the hubs of international trade. One can say that the first wave of globalisation took place several centuries ago, when cities utilised the increasing division of labour and transformed trade and related services into flourishing industries. They exchanged not only goods and materials but also cultural experiences. This should be borne in mind when globalisation, which for some has ominous connotations, is viewed in narrow economic terms. The popular conception that some multinational companies could exclusively manage a superior global network does not entirely match the reality. We also have political, social and cultural globalisation. Cities are the arenas where this all-embracing exchange takes place. Particularly in this new era of information technology, cities have developed a unique culture, for they have been historically prepared. They have now emerged as centres of communication, innovation and network building. Cities are participating in globalisation in its widest sense and, at the same time, are the site where this process takes place. In this respect, they can also be seen as the catalyst for every kind of progress in their respective countries.

From another perspective, cities in themselves are globalised. Places where several diverse communities live in close proximity appear to be microcosms of the world. This suggests that the "global village" is not simply a vision but already exists wherever cities emerge as multiethnic and multicultural entities. New York is the most convincing example of such a scenario but not the only one. This fascinating mega city reflects the globe in itself. This is true to some extent of many other big cities and is one reason why they are called metropolis. The other is that most of them over centuries have been frontrunners of globalisation – be it the ancient city states in Europe and around the Mediterranean Sea, colonial trade posts in Asia or pioneer towns in America.

Even today, big cities like Berlin, Cairo, Hamburg, Hong Kong, Jakarta, Kolkata (previously Calcutta), London, Mumbai (previously Bombay) Paris, San Francisco, Sao Paulo, Singapore, Sydney, and Vancouver have a lot in common: shape, structure, image, sensibility and burning issues. Most of them have gone through a process of chaotic urbanisation, digested waves of immigration, faced destruction by war and urban planning too, and experienced the cyclical ups and downs of prosperity and recession.

Furthermore, big cities broadly face similar problems. All of them have to deal with unpredictable demographic changes. They all try hard to compensate for the disappearance of ailing old industries by supporting new and more suitable branches of production. Urban economies are distorted by the simultaneous existence of widespread unemployment and shortage of professionals in certain fields. No big city has sufficiently overcome the momentous problem of burgeoning road traffic. And nowhere has the notorious housing shortage been adequately addressed.

Cities in the West, as well as those in the developing world, are still subject to severe mutilation from violent urban development that neglects historic city layouts and traditional urban lifestyles. The new paradigm is the spatial separation of functions like residence, work, and recreation, which reached dogmatic proportions through the Charter of Athens in 1933. Once this principle became a guideline not only for suburban settlements but also for the reconstruction of cities themselves, it became the main source for dilapidation of downtown areas and urban sprawl. Meanwhile in some cities, billions have been invested in reverse reconstruction and restoration of historic structures in order to revitalise their urban identity and thus keep the cities alive.

Last but not the least, the increasingly debt-burdened cities desperately try to maintain badly needed social services. Declining revenues caused by the departure of middle class people towards wealthier suburbs is reciprocally matched by growing poverty among the lower classes remaining in downtown.

These obviously comparable problems of big cities do not arise only from scarce resources and the lack of appropriate management. Solutions need to be considered in a wider perspective. They require a new culture of local politics in the spirit of good governance and greater participation by citizens as well. Even the best political intentions cannot achieve anything sustainable if the citizens themselves are not involved in the process of decision making. In other words, good governance and citizen participation are two sides of the same coin.

Altogether, urban development today cannot be understood as an isolated issue limited to a particular city. It is as global a theme as protection of human rights, preservation of the environment, international security, to name a few. Thus, cities must actively engage in building an international network of communication. Such a channel should be aimed at exchanging views, comparing experiences, jointly working out projects, and harnessing social and cultural diversity to enhance productivity. This will enable cities to become the driving force for overcoming religious and ethnic barriers and take a multicultural shape. This is what globalisation in terms of civil society means.

Manfred Haack
India Representative
Friedrich Ebert Stiftung

New Delhi
3 January 2003

Acknowledgements

The present book is the outcome of the international conference on Urban Environment Management: Local Government and Community Action organized by Institute of Social Sciences (ISS) in collaboration with Friedrich Ebert Stiftung (FES) in Kolkata in November 2001. This publication would not have been possible without the sincere support of various people involved in the different stages leading up to its final production. I am grateful to the Conference Committee members – Dr George Mathew, Mr Manfred Haack, Prof Mohit Bhattacharya, Dr Pabitra Giri, Mr T K Banerjee and Dr Swadesh Bhattacharya – who spent valuable time with us working on the agenda and design of the conference. Ms Madhulika Mitra of ISS, who worked very hard with me to make the conference a great success, deserves special thanks. All the delegates from India and abroad who participated in the conference enriched the discourse on urban environment management with their very enlightened contribution. I would like to especially acknowledge FES and its India Representative, Mr Manfred Haack, for their financial support and collaboration in organizing the event. The Kolkota Metropolitan Development Authority also deserves special mention for its support.

I greatly acknowledge the contribution of Ms Aparna Das for her editorial assistance in putting together this book. She painstakingly read all the papers and provided useful suggestions. I wish to thank Ms Niti Anand for her thorough and meticulous copy-editing of the manuscript. Mr. P N Kuttappan and Joshy Jose of ISS did the word processing. Mr Kuttappan merits special thanks for preparing the layout and taking care of all the other pre-production technical aspects of this publication.

Archana Ghosh
Conference Convenor
Institute of Social Sciences

New Delhi
1 January 2003

Contents

Contributors

Mr. Asok Bhattacharya, Minister-in-Charge, Municipal Affairs & Urban Development Departments, Government of West Bengal, India

Dr. R.K. Pachauri, Director – General, Tata Energy Research Institute, New Delhi, India

Prof. Mohit Bhattacharya, Former Vice-Chancellor, Bardhaman University, West Bengal, India

Prof. Amitabh Kundu, Centre for the Study of Regional Development, Jawaharlal Nehru University, New Delhi, India

Prof. Nazrul Islam, Department of Geography and Environment, University of Dhaka and Honorary Chairman, Centre of Urban Studies, Dhaka, Bangladesh

Mr. Tapan Kumar Banerjee, Former Director, Institute of Local Government and Urban Studies, Government of West Bengal and Honorary Director, Council of Development Studies, Kolkata, India

Prof. T. M. Vinod Kumar, Professor, and Co-ordinator, Centre for Urban Studies, School of Planning and Architecture, New Delhi, India

Mr. Seigbert Heid, Chairman, Working Group on Municipal Policy, Friedrich Ebert Stiftung, Germany

Mr. Helmut Dedy, Executive Manager, German Federation of Cities and Communities, Bonn, Germany

Prof. Asim Dasgupta, Minister–in–Charge of Finance, Government of West Bengal, West Bengal, India

Ms. Banashree Banerjee, Consultant, APUSP Project, Visakhapatnam, Andhra Pradesh, India

Mr. Lalith A Mallawarachchi, Colombo, Sri Lanka

Dr.L.A.S.Ranjith Perera, Assistant Professor, Urban Environmental Management Programme Asian Institute of Technology, Bangkok, Thailand

Prof. Golam Rahman, Department of Urban and Regional Planning, Bangladesh University of Engineering and Technology, Dhaka, Bangladesh

Ms. Renu Khosla, Associate Professor, National Institute of Urban Affairs, New Delhi, India

Mr. Chetan Vaidya, Principal Urban Adviser, Indo-U.S. FIRE Project, New Delhi, India

Ms. Sumona Bandyopadhyay, Lecturer, Department of Geography and Environment Management, Vidyasagar University, Midinapur, West Bengal, India

Dr. N. Ashok Kumar, Associate Professor, Regional Centre for Urban and Environmental Studies, Osmania University, Hyderabad, Andhra Pradesh, India.

1

Urban Environment Management: Local Government and Community Action

An Introduction

Archana Ghosh

The Context

In the last few decades, urbanization has become a powerful force throughout the world, especially in developing countries. It has been suggested that "while the whole world will become predominantly urban sometime between 2000 and 2010, the developing world will become predominantly urban by about 2015" (Stren 1994, 3). The growth of new forms of settlements and cities of unprecedented size has been a special phenomenon in the urbanization pattern of developing countries. According to UNCHS (1996), the average population of the world's 100 largest cities was over 5 million in1990 compared to 2.1 million in 1950 and less than 2,00,000 in 1800. The last few decades have also seen a world with a much higher proportion of the population living in metropolitan cities. The regional distribution of the world's total and urban population reflects the dominance of the million-plus cities in all regions, with Asia occupying the first position. Of the total world urban population of 5,28 million in 1990, Asia's share was 3,186 million or 32 per cent of the total population in the region; 42 per cent of the world's million-plus cities, including five(Tokyo, Shanghai, Beijing Mumbai and Kolkata), of the 10 largest urban agglomerations were situated in Asia (Ibid, 75).

The problem of accelerated growth of population in large urban centres has been compounded by the extremely limited financial resources of the developing countries. This has put a severe strain on their already weak and inadequate basic infrastructure, resulting in deterioration in all aspects of environmental quality at the household, neighbourhood and city levels. These manifest themselves in limited and unequal access to civic needs, such as appropriate shelter, sanitary toilets, safe water, sewerage and solid waste disposal systems, which inevitably creates health hazards. Even though in the last two decades, some two billion people in developing countries have gained access to better water supplies and another 400 million have got better sanitation facilities, these gains have to a large extent been offset by the enormous growth in the population of these regions. One billion people still do not have access to adequate supplies of water and two billion do not have proper sanitation facilities.

Not only are there inter-regional and intercity disparities in access to environmental infrastructure and basic amenities, differences are also sharply rising in micro-level environmental quality within a city — between high-income localities and slum and squatter settlements. Since large cities contain a very high proportion of these inadequate settlements, the disparity is most evident in these cities. As the environmental quality within cities deteriorates, the worst affected are the poorer sections. They lack the financial resources to compete for serviced land and adequate housing and are often compelled to occupy illegal settlements in unserviced lands along railway tracks, ecologically sensitive and hazard-prone areas like river slopes. The fragile environment gets further polluted with open defecation and indiscriminate garbage dumping in and around water sources. In India, on an average more than 40 per cent of the population in metropolitan cities lives in slums and squatter settlements situated in these vulnerable locations.

Environmental concerns of the cities of the developing world have in recent years hit the headlines, especially with a sudden spurt in diseases of epidemic proportions that were assumed to have been eradicated from the world long ago. The return of a cholera epidemic in Latin America in 1991 and the outbreak of plague in Surat in India in 1994 are instances of the serious risks posed to people's health by the lack of attention to basic environmental services like water supply, sanitation, drainage and solid waste collection.

People's awareness about environmental problems, and their impact on human health, and citizens' demand for according higher priority to environmental issues have increasingly attracted more attention from policy makers over the last decade. The ideal of "sustainable development" which stresses the need for "simultaneous implementation of both development and environmental goals" and "meets the needs of the present generation without compromising the ability of future generations to meet their own needs" (Brundtland Commission 1987) is embedded in the formulation of policy and action plans of all development programmes at the international and national levels.

A participatory approach among all stakeholders, with a direct and enhanced role of local governments, is the main emphasis of the sustainable development agenda for urban environmental management that has been endorsed in all international forums in recent years. It has been argued that local governments are the best protectors of the local environment. This is because they are the ones who construct, operate and maintain economic, social and environmental infrastructure, oversee planning processes, and establish and assist in implementing national and subnational policies. The Local Agenda 21 (LA 21) in Chapter 28 of Agenda 21 of the Earth Summit, or the United Nations Conference on Environment and Development (UNCED), in Rio in 1992 emphasized that the participation and cooperation of local authorities would be a determining factor in fulfilling the sustainable development objectives. The local authorities are to play a catalytic role in bringing together all the stakeholders, including the business community, government and non-government organizations, community organizations, youth, women and other marginalized classes, to make the city development process more inclusive. The agenda adopted at the second UN Conference on Human Settlements or Habitat II in 1996 also upheld the role of local governments in evolving a participatory approach with to include all stakeholders, towards fulfilling the goal of sustainable development. To address the complex problems of urban environment management, the Habitat Agenda stressed that effective decentralization of responsibilities to local authorities that are closest to and most representative of urban constituencies is imperative. Local authorities have thus been assigned an all-important role in the formulation and implementation of environmental action plans.

In the participatory approach to sustainable development, the role of the local community in need assessment and infrastructure planning is also widely acknowledged by Agenda 21 and Habitat II. It was earlier assumed that communities did not know their infrastructure needs; nor understood the environmental implications of not receiving adequate services; nor were they aware of the health-environment interlinkages. This myth was sedulously nurtured to legitimize top-down expert-led micro-level planning. It is now acknowledged that such an approach to need assessment and service delivery was one of the major reasons for the failure of many costly initiatives. Community participation is expected to be a component of all environmental infrastructure planning and decision making. Therefore, "What is clear is, first, that municipalities all over the world are now expected to undertake and to finance a much wider and more inclusive range of services and other economic and social activities. And, second, that the 'community' and important local stakeholder groups are pressing to be engaged in the local governance process"(Stren 2002).

Agenda 21 and Habitat II have definitely enhanced the roles of the local government and community in environment management. The available document on a cross-country survey on the implementation of LA 21 suggests that in developed countries, local authorities on a large scale have adopted its strategies even if they are not deemed as such. According to the survey, "In Western Europe new multi-stakeholder mechanisms are being implemented. Some 1,300 European local authorities, representing over 100 million citizens, have joined the European Sustainable Cities and Towns Campaign, committing themselves to engaging in Local Agenda 21 processes. All Swedish municipalities and over 90 per cent of the local governments in the United Kingdom of Great Britain and Northern Ireland have adopted these strategies. In Australia 177 local governments have adopted local Agenda 21 or similar sustainability strategies" (ICLEI, 2002).

However, there is considerable variation in the number of local authorities across countries belonging to the developing world that are engaged in, capable of, or even empowered to take up participatory environmental planning and management at the local level. This is so notwithstanding the fact that the goal of sustainable development via the implementation of LA 21 have been adopted by almost all countries. Examples of putting into practice a participatory approach in managing the urban environment are still not many in developing countries. Evidence suggests that, "... Local Agenda 21 planning currently is more prevalent in developed countries"(ICLEI 1997). The hindrances to the development of effective participatory processes could be many. These could range from the lack of a proper institutional, legal and financial framework at the national, local and community levels; to policies, procedures and rules not being in place to provide an enabling environment; to the absence of appropriate programmes to build the capacity of all stakeholders, including local governments, NGOs, community organizations, private sectors players and others.

However, there is a positive trend towards more participatory approaches in many cities of the developing and least developed countries, even if they do not strictly fall within the parameters of LA 21. Local authorities are adopting several measures – financial, administrative and technological – related to development planning – to improve the quality of living in cities; community organizations and NGOs are joining hands with local governments to mobilize funds for their water and sanitation schemes or to operate and maintain the existing ones. Even though the success stories are not very large in number and the progress is slow there is a definite trend towards a participatory development agenda in cities around the world, including those in South Asia.

The Conference

The international conference on Urban Environment Management: Local Government and Community Action was organized by the Institute of Social Sciences (ISS) in collaboration with the Friedrich Ebert Stiftung (FES) on 25-27 November in Kolkata to take stock of the ideas generated and experiences gathered so far in the above area of global concern. Kolkata, the capital city of West Bengal in the eastern part of India, was selected as the venue for several reasons. Kolkata is one of the largest, most densely populated and environmentally most sensitive mega-cities of the world. The Kolkata Municipal Corporation (KMC) has a strong political culture and long history of democratic governance through the Mayor-in-Council(Cabinet form of local government) system. The state government and KMC, in recent years, have taken several new steps to address environmental problems via a participatory infrastructure delivery system, where the private sector, NGOs and CBOs all play important roles.

The conference was held at a time when the world was preparing for the second Earth Summit, or Rio plus Ten, in Johannesburg, South Africa, scheduled to be held in August-September 2002. The conference provided an appropriate forum for discussing the issues of sustainable development and environment management and to assess how far the world has progressed on the path to attaining the objectives set by the first Earth Summit in 1992.

The conference also provided the participants with the opportunity to share their experiences and set the agenda for the future. As Manfred Haack, the India Representative of FES, said in his address in the inaugural session, the world's biggest cities themselves are globalized, attracting people from across the globe. With the recent spurt in the globalization of the economies of various countries around the world, the need and opportunity for networking among cities to foster social and cultural ties has increased. All the major cities of the world experience similar kinds of problems in managing the economy, environment and civic amenities. People from different parts of the world can learn from each other's experiences and exchange ideas to solve problems through good governance and citizens' participation. The conference provided a forum for such networking and interaction.

The Objectives

The specific objectives of the deliberations of the conference were primarily to: *a*) discuss the extent and nature of environmental and related problems faced by the cities and towns in the countries of South and South-East Asia; *b*) understand how local governments, community groups and NGOs are repositioning themselves to meet the emerging environmental challenges; c) to learn about the success stories and models of urban local government and community intervention in urban environment management; and d) recommend some measures for the future.

The format of the conference was as follows: presentation of the main themes by eminent experts in the first plenary session; working group meetings on issues related to the main themes; and open house discussions in the subsequent plenaries. The three main themes of the conference were:

1. Environmental implications of rapid urbanization and the poverty-environment interface
2. Issues in environment management and the role of local governments
3. NGO and community initiatives in environment management at the settlement level.

During the three working group meetings and the subsequent open house plenaries, there were more in-depth discussions on critical aspects of the following issues:

- ❑ Environmental implications of inadequate services
- ❑ Institutional, fiscal and legal issues, with special reference to role of urban local governments in participatory environment management
- ❑ Prospects and problems of community and NGO participation in environmental management

Around 200 urban experts, the mayors of the cities of Colombo, Karachi, Kolkata and a few other Indian cities, and local government and NGO representatives from Bangladesh, Pakistan, Sri Lanka, Thailand, Nepal and Germany and different parts of India extensively participated in the three-day deliberations. Political leaders, NGOs, academics and officials from the concerned government departments of West Bengal and the host city Kolkata made rich contributions to the discussion from their experiences in managing the city's environmental problems. Though the focus of the conference was South and South-East Asia, scholars from Germany and representatives of several donor agencies, such as World Bank, DFID and UNICEF enriched the discussion with their respective perspectives.

The Deliberations

The conference deliberations were rich in content and context. On the one hand, concern was expressed by almost all the speakers about the deteriorating environmental conditions with increasing urbanization in South Asian countries, threatening the sustainability of their cities and towns. Strong apprehensions were expressed that the mere availability of financial resources, technological know-how and administrative skills were becoming increasingly inadequate to manage cities today. On the other hand, there were presentations on new models and success stories of local government initiatives; collaborative efforts adopted at the city level in different countries; and people's initiatives at the grass-roots level for solving environmental problems, which aroused hopes for the future.

Concerns

Delegates from Bangladesh, Sri Lanka, Nepal and India -all presented a grim picture of environmental services in the cities of their respective countries. Almost all the cities in the region suffer from specific environmental problems related to poor waste water treatment; inadequate solid waste collection and disposal; discharge of waste water in rivers and other water sources; growing pollution from ever-rising vehicular traffic; and proliferation of slums and squatter settlements. Growing air and water pollution and their impact on citizens' health were the common concerns. Keynote speaker R K Pachauri's estimates that India is losing 10 per cent of its GDP on account of environmental pollution, and that around 2.5 million people in the country die prematurely each year as a result of air pollution, were extremely disturbing. It was observed that both absolute poverty and the gap between the rich and the poor, in terms of shelter and basic environmental services, were increasing with the spread of globalisation and the market economy. Environmental legislation is in place in many countries, the proper institutional structure for strict enforcement is lacking.

The problems of European cities are different from those in the developing world. Helmut Dedy has argued in his paper that the dispersion of urban settlements into monofunctional districts, based on the division of labour, and the resulting growth of traffic are creating environmental problems in German cities.

The role of local governments in meeting citizens' needs were cited by many speakers as crucial, especially in the decentralized political system advocated and adopted in many developing countries such as India. The lack of sufficient resources with local governments to manage urban infrastructure development was a recurring concern expressed in various papers and in the open forums. Internal financial management problems and not-so-perfect accounting systems of the local bodies highlighted by T K Banerjee, Aswajit Singh and other speakers during discussions, are serious stumbling blocks in effective resource planning for them even though they have the potential to attain self-sufficiency in resource mobilization. The lack of political initiatives for internal financial and administrative reforms and the lack of transparency and accountability account for their poor performance.

The observations in this regard by the mayors of cities, be it Karachi, Kolkata or Colombo, were almost identical. Even though some new initiatives for resource generation are being taken by local bodies in different countries, the results are neither very encouraging nor replicable by all types of local bodies and in different national contexts. Amitabh Kundu expressed reservations about the recent trends towards economic liberalization and political decentralization, which are seen as the harbingers of sustainable development. He pointed out that there are ample reasons to be skeptical about the capacity of municipal governments to meet the needs of the micro-environment. According to him, the new methods for municipal resource mobilization, like municipal bonds, adopted by some municipal corporations in the metropolitan cities of India, are more burdensome for city governments. Further, these methods cannot be adopted by the smaller municipalities at all because of their poor financial conditions. They will only deepen the divide between cities of different sizes in their access to basic services and widen the gap between rich and poor city-dwellers. Participants also expressed concern over the withdrawal of the public sector from infrastructure development in cities. Since city governments are perennially short of resources, urban sector reforms in developing countries are encouraging private sector participation. This, in turn, is paving the way for the displacement and eviction of slum-dwellers and the poor in the name of development and environmental upgradation, from sites where they have lived for many years.

The question of subsidies repeatedly came up in many papers. Since the policies on subsidies are not transparent, subsidized municipal services in most cities in developing countries benefit those who can afford to pay while the poor remain deprived. In fact, the poor pay more for services than the average citizens because they do not have adequate access to municipal services and have to resort to non-municipal sources. In contrast, as Dedy points out in his paper, in Germany, the access to services is not linked to subsidies; it is only determined by the ability to pay.

The market-driven paradigm adopted by developing countries in recent times is creating disparities between cities in terms of basic access to amenities and putting them on a competitive platform to attract foreign investment. An alternative picture emerges from Seigbert Heid's paper on Germany, where the financial equalization scheme adopted by the federal government for federal states and municipalities is trying to even out disparities in the quality of living between people in different parts of the country.

Multiple authorities engaged in city planning and development without a clear demarcation of their administrative jurisdiction pose a problem in all major cities – be it Karachi, Mumbai, Kolkata, Dhaka or Delhi. The newly elected Karachi Deputy Mayor expressed his helplessness over the fact that the Master Plan and Environmental Control Department of Karachi

Development Authority is responsible for the preparation of plans but implementation is divided among different city agencies, which have poor coordination among themselves. Mohit Bhattacharya, R M Kapoor, K C Sivaramakrishnan and several other speakers from India spoke about the problems arising from the existence of multiple agencies. This issue has not been resolved till date despite the enactment of the 74th Constitutional Amendment in 1992, which empowers local governments to take up city planning and urban development.

The trend of devising metropolitan development programmes without giving due consideration to environmental planning is most alarming to environmentalists. Sumona Bandyopadhyay's paper on the newly developed township in Vasai-Virar subregion of Mumbai shows how the township has grown without long-term environmental planning. This has generated concerns about the sustenance of its wetland ecosystem and the inadequacy of basic services for its residents.

New Initiatives

The importance of the community's role in collaborative initiatives towards environmental planning and management for sustainable development strongly emerged during the conference deliberations. Even though the integrated approach suggested by Local Agenda 21 has been formally adopted by many countries – to ensure participatory city development strategies with an enhanced role for civil society and other partners and information sharing between all the stakeholders – very few city governments in Asia have put it into practice. Contrary to the trend suggested by the ICLEI survey as mentioned in this paper before some of the countries in developed world are also lagging behind in the implementation of LA 21. As Heid's paper points out that only 12 per cent of municipalities are actually implementing LA 21 in Germany.

However, there are several isolated examples of new initiatives being taken by local governments for urban development through participatory measures. The community is also directly or indirectly participating in the development process. Colombo Mayor Omar Kamil's presentation on Colombo Municipal Corporation's (CMC) very successful initiatives was an eye-opener not only for the newly elected young Mayor of a city like Karachi but for other city leaders as well. CMC has not only involved business communities and the private sector in the provision of urban infrastructure but also introduced professional management principles in city governance. The extensive involvement of the community through community development councils, voluntary participation of highly experienced city residents through city watch groups and the active participation of the business sector, NGOs and bilateral aid agencies have made environment management a truly participatory effort in Colombo. CMC was awarded the first prize at the 2000 Asia-Pacific Mayors' Conference Honolulu for meeting its environmental commitments.

Among the good governance principles for sustainable development is transparency and accountability. An example of this is found in the state of Maharashtra in India. B M Sukthankar in the open discussion talked about the recently introduced provision in the municipal legislation of the state. This provision has made it obligatory for municipal commissioners to put before the standing committee detailed accounts of income and expenditure, the gap between the two, how this gap is met, whether local bodies are giving subsidies for this purpose, and who are its beneficiaries.

Chetan Vaidya and Renu Khosla's paper on bankable infrastructure projects in the municipalities of Ludhiana and Sangli shows how community mobilization has enabled the water supply projects in these two cities to fulfill the twin objectives of providing water to the poor and cost recovery. The paper shatters the myth that the poor do not pay for municipal services.

Ranjit Perera's presentation on the strategic planning (SP) approach adopted by some city governments in Sri Lanka and Thailand brought out how cities are repositioning themselves to meet the immediate needs of capital formation in a competitive environment. Compared to the Master Plan and Structure Plan-based city development, which requires colossal amounts of money and long implementation periods, the SP is more flexible and prompt and is especially appropriate for medium cities. Its bottom-up approach ensures extensive stakeholder involvement. Asia's experience of participatory urban governance for environment management in Colombo Municipal Council in Sri Lanka and in Phuket and Nonthabari in Thailand are successful examples.

The Clean and Green Programme of Hyderabad city, also implemented in certain towns of Andhra Pradesh, is an example of a multisectoral and participatory programme for environment protection. The Andhra Pradesh Urban Services for the Poor (APUSP) project, on which a presentation was made by Banashree Banerjee, is a highly participatory environmental improvement programme for slums. APUSP follows a three-pronged strategy: (a). reform of municipalities to improve the performance of their poverty-reduction activities; (b). improvement of environmental infrastructure in slums; and, (c). strengthening of civil society to stimulate policies in favour of the poor and to facilitate interaction between them and municipalities.

Golam Rahman's paper examines the substantial role played by Waste Concern, in implementing community-based waste management practices in some parts of Dhaka city. The presentation on the Healthy Cities Programme of World Health Organization, which is being implemented in some cities of Bangladesh, in another example of a participatory environmental improvement programme at the local level.

Future Agenda

During the conference deliberations, several recommendations were made and guidelines suggested for future action. Both the inaugural address by Asok Bhattacharya and the keynote address by R K Pachauri reiterated that since the poor are the worst victims of environmental degradation, and they are compelled to migrate from rural to urban areas in search of livelihoods, concomitant rural development and poverty alleviation policies and programmes are necessary for sustainable urban development. Asim Dasgupta in his valedictory address added that effective and efficient urban environment management should include not only physical infrastructure development but also improvement in the social environment, including issues relating to livelihood, access to employment and education, and, most importantly, land ownership and tenure security.

Several delegates emphasized the need for collaborative efforts on the part of all stakeholders in urban environment management. They reiterated the need for a more rigorous implementation of LA 21 and the Habitat II agenda of decentralized urban governance and enhanced role of urban local bodies and community organizations in achieving the goal of sustainable urban development. Nazrul Islam pointed out that civil society can play an important role through advocacy programmes; and by keeping the pressure on both the central government and local governments, on the one hand, and the private sector, NGOs and CBOs on the other. This can arrest the environmental degradation and make cities more livable. Alison Barrett noted that for participatory processes to be adopted on a large scale, a clear institutional structure needs to be in place; otherwise, the participation will be ad hoc and dependent on the individuals involved. More importantly, in order to develop equal partnerships between local governments, NGOs, community groups and the private sector, there needs to be a change in mindset. A new culture must emerge, one where citizens must acknowledge their share of responsibility in both creating urban environmental problems and finding solutions to them.

The present publication is an effort to document the rich debate generated by the three-day conference during the presentation of theme papers and case studies and the highly interactive discussions that followed. It is hoped this volume will provoke more debates on the issues raised; motivate practitioners in this field to undertake further research; and provide an opportunity for all stakeholders involved in the task of urban environment management to learn from the failures, and be inspired by the success stories documented here.

REFERENCES

Brundtland Commission (World Commission on Environment and Development) 1987: *Our Common Future,* United Nations Environment Programme.

ICLEI (International Council for Local Environmental Initiatives) 1997, *A Study of Responses by Local Authorities and Their National and International Associations to Agenda 21,* ICLEI, Toronto.

------- 2002: *Accelerating Sustainable Development: Local Action Moves the World.* Report prepared for the Commission on Sustainable Development, acting as the Preparatory Committee for the World Summit on Sustainable Development 2002, ICLEI Toronto.

Stren Richard 1994: "Towards a Research Agenda for the 1990s: An Introduction" in *Urban Research in the Developing World: Asia,* Centre for Urban and Community Studies, University of Toronto, Toronto.

------- 2002, "Decentralization: False Start or New Dawn?" in *Habitat Debate,* (March) vol. 8, no.1. Newsletter for UN-Habitat, Nairobi.

UNCHS (United Nations Centre for Human Settlements) 1996, *An Urbanizing World: Global Report on Human Settlements,* Oxford University Press, Oxford.

2

Urban Development and Environment Management

Initiatives in West Bengal*

Asok Bhattacharya

I am extremely happy to be here this evening in the midst of this august assembly. I would like to congratulate the organizers for choosing a very relevant theme for the conference. I also wish to thank them for holding this international conference in Kolkata – one of the large metropolises in the world beset with problems pertaining to environmental management.

The government of West Bengal has envisioned the provision of a healthy and sustainable living environment for residents in the urban areas of the state. Before this vision can be translated into reality various challenges need to be met, particularly in a developing economy like ours. We are aware of the global concern about environmental sustainability, culminating in Agenda 21 of the Earth Summit in Rio de Janeiro in 1992. However, the implementation of this agenda is a complicated task.

As a minister in charge of both urban development and municipal affairs for about a decade, I have had the opportunity of working closely on various issues concerning environmental sustainability. From my experience I have found that no environment management programme in cities and towns can be sustained without the participation of the local community as also of the elected local government. This can happen where a decentralized model of development is adopted. In my view, the existence of poverty is the greatest impediment to providing a

---

* Inaugural Address.

healthy and sustainable environment and decentralized democratic governance can effectively respond to the issue of poverty reduction. Poverty is a multidimensional concept encompassing economic, social, political and cultural aspects. The devolution of resources and certain functions and decision-making powers to democratically elected urban local bodies (ULBs) as a consequence of the policy of decentralization can aim to reduce poverty through an *empowerment* strategy. This strategy creates "space" for people to effectively participate in the decision–making process, including the setting of priorities for budgetary allocation. Indisputably, local-level needs and priorities cannot be reflected in the investment programme and other actions unless ULBs are directly involved in the design of such interventions. Again, without the participation of local communities, ULBs cannot do justice to their work. Further, poverty reduction and decentralization can be linked through a resource mobilization strategy with local people committing more resources to UBLs over which they are likely to have greater control. There is evidence to suggest that ULBs can gradually develop a comparative advantage in the promotion of effective poverty reduction strategies by virtue of their being able to have vertical and horizontal information and insights and to coordinate mapping and mobilization of local capacities and resources. Often, development potential remains dormant at the local level. This can be best identified and adequately tapped by ULBs, in association with local communities, for local economic development. The involvement of local communities, particularly women, in project design and implementation develops a sense of identification with the assets created. This leads to improved management of infrastructure and services and, consequently, effective sustenance of the benefits of such interventions. Therefore, without community participation, the decentralized paradigm is not likely to be successful and, as such, management of the urban environment is unlikely to be sustained.

It would be unfair on my part if I did not mention two other policies of the state government that have bearing on a healthy and sustainable urban development in West Bengal. The state government has rightly appreciated that many of the urban problems have roots in rural areas and, as such, there is a need for integrating rural and urban development. With this in view, the state government has assigned high priority to agrarian development and carried out a number of reforms in rural sectors. As a consequence, there has been significant growth and prosperity in the agricultural sector, which has put a brake on migration from rural to urban areas, particularly to Kolkata Metropolitan Area (KMA). The impact of this on urban development can be easily understood. The policy of decentralization apart, agrarian prosperity has created growth potential in small and medium towns spread over various districts in the state. The other policy parameter pertains to the emphasis placed by the state government on the development of small and medium towns across the state.

The state government adopted the decentralized approach to urban planning and development as early as the beginning of the 1980s. This new paradigm was implemented, in the first instance, in the third phase of the World Bank-assisted Calcutta Urban Development Programme (CUDP-III), which was then being formulated.

The government was aware that in order to operationalize the decentralized development approach, it was necessary to introduce certain legislation and institutional reforms were necessary. Towards this end the following steps were taken:

- Enactment of Kolkata Municipal Corporation Act and Howrah Municipal Corporation Act 1980 replacing the Calcutta Municipal Act of 1951 to provide for a cabinet form of local self- government
- Overhauling of the Bengal Municipal Act of 1932 culminating in the West Bengal Municipal Act 1993
- Establishment of the Directorate of Local Bodies (DLB) and the Municipal Engineering Directorate (MED) at the state level, and of Municipal Development Programme (MDP) Directorate within Kolkata Metropolitan Development Authority (KMDA, previously CMDA) to extend the necessary assistance, technical and otherwise, to ULBs

- Setting up of Institute of Local Government and Urban Studies (ILGUS) for conducting certain basic studies on urban issues, on the one hand; and imparting training for upgradation of human resource skills, on the other
- Setting up of Municipal Finance Commission (MFC) to inquire into the fiscal health of ULBs and to recommend necessary measures for the fiscal strengthening of ULBs.

CUDP–III included the MDP, under which investments were earmarked for each of the 41 constituent ULBs of KMA. Each ULB was required to formulate an investment plan for the development of different sectors of urban infrastructure and services based on local needs and priorities. The ULBs were required to provide an undertaking to the effect that they would operate and maintain the infrastructure on their own. Annual fiscal targets were set for each ULB pertaining to sustained improvement of the assets created. Beside MDP, two other major components of CUDP–III were the Small Scale Enterprises Programme (SSEP) and the health programme specifically meant for the reduction of urban poverty.

We in the state government have always understood that reduction of urban poverty – through the generation of income and employment opportunities among the urban poor, coupled with the provision of social infrastructure and services like primary health, skill upgradation, mother and child health, legal literacy of women, etc. – will play a key role bringing about sustained improvement in the people's living environment. Besides SSEP and the health programme under CUDP–III mentioned above, two major projects of this nature in KMA that are currently being implemented are the Calcutta (Kolkata) Slum Improvement Project (CSIP), assisted by Department for International Development (DFID) of UK, and the India Population Project (IPP)-VIII, assisted by the World Bank. Further, the ULBs, particularly those outside KMDA, are also implementing a number of schemes sponsored by the Government of India, such as Swarna Jayanti Shahari Rojgar Yojona (SJSRY), the National Slum Development Programme (NSDP) and others. The significant and common approach of these projects has been the involvement of the local community at the grass-roots level. In some cases, NGOs have also been involved. The evaluations of CSIP and IPP–VIII have shown that the active participation of the local community in planning, design and implementation have rendered the projects successful and sustainable. The approach and strategy followed by both the projects led to a remarkable improvement in maternal and child health and a reduction in fertility rate in the KMA. It has been observed that the crude birth rate registered a decline from 19.63 in 1992 to12.78 in 1998. The infant mortality rate and the maternal mortality rate came down substantially – from 55.58 and 4.56 to 22.7 and 0.3, respectively – during the same period. Now there is a need to link the micro-organizational structure with the elected ULBs and also with the nodal agency, such as KMDA, at a higher level.

The state government's initiative on a decentralized paradigm of planning and development received legislative support from the 73rd and 74th Constitutional Amendment Acts of 1992. The West Bengal government enacted the District Planning Committee (DPC) Act and also the Metropolitan Planning Committee (MPC) Act according to the mandate of these constitutional amendments. The government has issued orders for establishment of a ward committee in each ward of a city/town, irrespective of its population size. The government has set up the West Bengal State Finance Commission to determine the allocation of revenues between the state and local governments. West Bengal is the first state to have constituted a Metropolitan Planning Committee (MPC) for KMA in 2001. The growing trend towards decentralization, visible during the last two decades or so, has led to a healthy working relationship between KMDA and ULBs within KMA.

The continuing concern of the state government for improvement in the management of the urban environment, in association with ULBs and local communities, can be discerned from the two major projects that are awaiting implementation shortly. The Kolkata Environment Improvement Programme (KEIP) is to be taken up by Kolkata Municipal Corporation (KMC), with assistance from Asian Development Bank. The programme will have two major components – namely, capacity strengthening of KMC and certain infrastructure projects that have direct bearing on environmental improvement. Under the latter, there shall be schemes for sewerage and drainage, solid west management (SWM) system improvement and the development of selected slums. Kolkata Urban Services for the Poor (KUSP) is being designed under the aegis of KMDA, with assistance from DFID of UK. The major objective of this project is to improve the quality of life of the poor in KMA. The project will be planned and implemented by KMDA, with the active participation of ULBs and other community based-organizations. Infrastructure development under this project will be taken up with a view to effecting improved delivery of basic services to the urban poor.

In appreciation of the need to have a long-term development perspective and to formulate sectoral master plans for different sectors of urban infrastructure so as to strike the right balance between development and environmental considerations, KMDA has prepared Vision 2025. Essentially, this is a development perspective, covering the next 25 years, for KMA, including the Master Plan for Traffic & Transportation for KMA. Vision 2025 has suggested various strategies to provide a healthy living environment for the projected population of KMA. The Master Plan has suggested various projects that should be taken up over the next 25 years to cope with the growing volume of traffic – projected to rise from one million to three million – in an environment-friendly manner.

Notwithstanding the successes achieved, various shortcomings still remain. There is yet the necessity of strengthening the fiscal capability of ULBs. In this context, the need for resorting to cost recoveries against certain municipal services, such as, water supply and solid waste management, cannot be overemphasized. Upgradation of the technical capabilities of ULBs – in terms of skill augmentation of municipal personnel and introduction of IT-based information assembly and networking in each ULB – are considered imperative for the effective functioning of ULBs across the state.

In my opinion, the gravest concern pertains to poverty that is inextricably related to economic growth, particularly of non-agricultural activities in urban areas. The process of globalization unleashed by the constitution of the World Trade Organization (WTO) has created a unipolar world in which developing countries like India are engaged in an unfair competition with the developed countries. Evidently, there are signs of industrial decline in developing countries, including India. It is doubtful if any poverty-reduction strategy can be successful in such a context. It needs no prescience to suggest that environmental sustainability without poverty-reduction is not likely to be achievable. It is this concern that informs the priority accorded by the West Bengal government to such projects and schemes that directly address the urban poor, especially with respect to generation of income and employment.

I believe that the deliberations during this conference will deal with issues that are both local and global in nature; and that useful recommendations will be made that may benefit governments in formulating effective polices for addressing environmental management issues. On this note of cautious optimism, I inaugurate the conference and extend my good wishes for its success.

3

Issues in Urban Environmental Management

Agenda for Change*

R K Pachauri

At the outset, I would like to express my gratitude for being given this opportunity to address such a distinguished group of participants and be part of such an eminent panel of speakers at this inaugural session. If I may be permitted to continue on a personal note, let me share with you a little bit of information about my institute. It is called TERI (Tata Energy Research Institute). It is not a Tata organization. We were given the seed money by the Tata group to get started, but we are a totally independent non-profit institution. Energy is a small and important part of the work that we do. Our other areas of work include environment, climate change, biotechnology, forestry and a host of other subjects. Besides research, we are also involved in a great deal of field activity and action-oriented work, which I shall refer to as I go along. Finally, let me also say that we are privileged to have as chairman of our governing council Dr. Arcot Ramachandran, who is an acknowledged authority on urban issues who used to be the head of UN-Habitat for a number of years.

The Hon'ble Minister and the Hon'ble Mayor of Kolkata have already said a great deal on what I wanted to talk about. But I am delighted to find that in spite of my lack of experience, I am in tune with what they essentially culled out after years of experience. Let me emphasize that in a society such as ours,

---

* Keynote Address

urban environmental management (UEM) essentially has to begin with enlightened rural policies. The Hon'ble Minister made the point that there are two major reasons why this statement can be treated as axiomatic. First, it is the economic impoverishment and absence of proper employment opportunities that drives the huge volume of migration from rural to urban areas, particularly into large cities in the developing world. The result is that the most underprivileged, the most unskilled and, therefore, the most helpless individuals constitute the largest share of migrants from rural into urban areas – who move in the faint hope of escape from their endemic grinding poverty. And poverty, we need to be reminded, is not merely a function of low income. It is a combination of various factors that amount to a lack of opportunities and the absence of social empowerment. Referring to Prof. Amartya Sen's ideas on poverty, it is deprivation of basic capabilities rather than merely low income, which can be reflected in premature mortality, significant undernourishment, especially of children, persistent morbidity, widespread illiteracy and other failures. Unfortunately in this country, we have not been able to create opportunities and avenues for local empowerment in rural areas. Though I must say that the record of West Bengal has been better than that of other states in general. The second reason why rural policies often decide the state of the urban environment is to do with the fact that in very basic activities related to cooking and hygiene, rural migrants carry established habits and practices into urban locations as well. It is not unusual to find that in urban slums, cooking is carried out with the use of poor quality biomass, even in cases where slum-dwellers may have income levels that make superior fuels affordable. Indoor air pollution in our country is a far more serious threat than outdoor pollution. TERI carried out a major project called Green India – 2047, which established that approximately 2.5 million people die prematurely in this country each year as a result of air pollution. A major component of this is indoor air pollution. Women and children are exposed to long hours of indoor pollution as they sit before cooking stoves of very poor design, using fuels of very poor quality in homes that have very poor ventilation. Similarly, in the areas of personal and social hygiene, sanitary habits seldom change when persons move from rural to urban areas.

A basic reason why the urban environment has steadily deteriorated during the past two to three decades is motorization and a massive increase in urban transport. This, of course, stems from a sharp growth in the country's urban population leading to activities which are directly in proportion to the density of population in a given area. A prime example of this staggering growth is the city of Delhi. Delhi had 4 lakh persons in 1901. In 1961, its population was 2.65 million; by 1991, it had risen to 9.42 million. In 2001, including the greater Delhi area, the population is around 14 million. The decadal growth rate during 1981-91 was about 51.45 per cent and during 1991–2001 a little lower at 46.31 per cent. Such staggering growth will obviously stress the environment and all forms of municipal services. In the case of other cities – be it Bangalore, Hyderabad or Ludhiana – the enormous demand for services and limitations of space invariably leave the urban environment heavily stressed. The footprints of urban activities on the natural resources of various cities, are therefore growing very rapidly. Thus far, urban bodies have not been able to develop adequate services for the population – with important exceptions of course; – apart from finance they also lack the organizational capacity to provide urban services in requisite measure to keep pace with rapidly growing demands.

The irrational pricing of services only compounds these problems, as was referred to by the Hon'ble Minister. He cited the example of pricing of water, where incentives for efficient use of water do not exist in the pricing schemes offered to consumers. In fact, underpricing of water, irrespective of the type of usage and status of the user, only encourages gross wastage. I am delighted to know that in the project that is going to be supported by the Asian Development Bank for Kolkata's environmental improvement one of the major components is stakeholders' consultative process (SCP). I believe that through this process of consultation, we can convince the users that payment for the services would ensure efficient and better services. There is no reason why consumers will not be open to such ideas.

It would not serve any purpose to project the staggering increase in demand for services that would be faced in urban areas in the country in the coming decades. Inherent in this expansion is the rapid deterioration of the urban environment. Perhaps one single area illustrates the enormity of the challenge. If we look at the problem of solid waste management for the country as a whole, we find that more than 25 per cent of the municipal solid waste generated everyday in the country is not collected at all and therefore becomes a perennial source of pollution. Almost 70 per cent of Indian cities lack adequate capacity to transport solid waste; the sanitary landfills for disposing off the waste are neither well-equipped nor well-managed. Our own estimate is that in 1947, cities and towns in India generated an estimated 6 million tonnes of solid waste. Fifty years later, this figure had risen to 48 million tonnes. In the next 50 years, the figure will be in the neighbourhood of 300 million tonnes. If we continue with the current trend and practices, we will need more than 170 sq. km for landfill sites alone. With regard to treatment, of the total quantity of over 30 million liters of sewage generated a day, barely 10 million liters are treated in all – and that too in an ineffective manner. This is because sewage treatment facilities even where they exist, function poorly, often characterized by a state of disrepair, inappropriate design, poor maintenance and ineffective management.

If the quality of water in our major rivers is to be improved, effective sewage treatment is an imperative. There are 113 river basins in India, of which 14 are major, 44 are medium and the remaining 55 are minor. The major river basins are those of Brahmaputra, Ganga, Indus, Godavari, Mahanadi, Narmada, Cauvery, Brahmani, Tapti, Mahi, Subarnarekha, Pennar and Sabarmati. Water in most of these suffer from severe problems of quality despite the fact that we are monitoring the quality of our rivers. There are 480 stations under different programmes, where several monitoring parameters are used – for assessment of water quality and quantity, including physical, chemical and biological. Of these, biochemical oxygen demand (BOD), dissolved oxygen and chloroform count are perhaps the most important. Sadly, the worst polluted river in the country is the one that the father of the nation, Mahatma Gandhi, was associated with – the Sabarmati river following through Gujarat. Its BOD values in 1997 were greater than 20 milligrams per liter, which is substantially above the permissible limit. Even in Ganga, the BOD exceeds the levels for class C quality, which applies to water fit for drinking after conventional treatment. Clearly, this is symptomatic of the problem that we have with water quality. Unfortunately, the cost to society of this deterioration in water quality is never estimated. Here I would like to propose the need for confluence of environmental action and good economics. In the case of water quality, if we were to take into account the loss of fish and other living organisms in our rivers, the cost of treatment for making water fit for drinking and bathing and other human uses, as well as the loss of recreational values, we will find that the overall cost of poor water quality would be staggering. The country is losing far more in economic terms than it would cost to treat sewage and other effluents being discharged into our river systems. Sewage discharge is, of course, compounded by the discharge of pesticides, chemical fertilizers and other chemicals from industries. This is what makes the clean-up operations very difficult.

Now I want to briefly refer to the branch of environmental economics called natural resource accounting, which essentially applies economics methodology for estimating the costs and benefits of natural resource management decisions. This is where our economic policies have been largely deficient – by laying sole emphasis on national accounts, both at the national and sub-national levels without taking into consideration estimates based on natural resource income accounting. We have not applied this system of accounting effectively. Under the Green India 2047 project carried out by our institute, we estimated that over 10 per cent of India's GDP is being lost on account of environmental damage from degradation of natural resources and related impacts. Urban systems therefore have to be improved so that they do not exert parasitic pressure on the natural resources of the country. Now my purpose in giving these figures is not to paint a doomsday scenario. It is only to highlight the huge magnitude of the challenge before us and to emphasize that a change in direction is required. What are the changes that we need to make in order to solve the problems facing us? I will list the following four major initiatives for the consideration of this enlightened audience.

Agenda for Change

1. Environmental Performance Assessment

A system of assessing environment performance needs to be introduced in our urban areas to monitor the services provided and their overall effectiveness. This information should be provided to the public so that they can judge whether their tax money is being properly utilized or not. This is where modern technology can make a major difference. People should have easy access to this information, either by putting it on a website that is easily accessible or via regular mails to organizations. A system of performance measurement and monitoring, as well as dissemination of information related to it, will also create healthy competition between municipal bodies to improve their services. Most importantly, an objective method of continuous assessment and total transparency will ensure greater community participation and action by civil society. This is where the SCP becomes extremely important. TERI is carrying out a major project, supported by the Ministry of Environment and Forests and UNDP, whereby the Institute has developed a methodology for assessing environmental services and rating different municipal bodies on how effectively they deliver these services. This information will be made available to the public and a dialogue with the stakeholders will be conducted so that they are fully aware of the performance of the bodies that are responsible to them. With performance measurement and monitoring must also come benchmarking. This has to be based on sound practice. Unless you have benchmarks, mere comparisons only let you learn of inferior services, which does not serve the purpose.

2. *Regulation of Urban Services*

There is a need for independent regulation of urban services with the full involvement of the community. This will enable the proper pricing of services, an assessment of the efficiency levels and the development of a framework whereby private sector investments can be attracted. The private sector has to be given the assurance that the investments made and resources employed by them will earn them a fair return, but they have to be monitored by independent regulatory bodies. Given the fact that several sections of society have the capacity to pay full user charges for the services they receive, the pricing system will have to be modified to ensure fairness and equity for all stakeholders and efficiency in the supply and use of services provided. There is no reason why a service such as solid waste management cannot be privatized. Private sector operators can create partnerships with NGOs and communities for the collection and segregation of wastes, part of which could be used for generating energy through the bio-methanation method. Similarly, private sector initiatives are possible in the management of water in urban areas. Funds can be obtained for projects of this nature. It requires appropriate designing of the projects and submitting proposals to the right funding organizations. The concerns about the poorer sections of society being deprived of the services and their inability to pay for them can also be addressed. This can be done either through direct subsidies, which can be financed by tax collections and other revenues, or through cross–subsidization by the richer sections of society. This is where the right government policy and adequate regulation can protect the interests of all citizens in a particular city.

3. *Environmental Improvement Plan*

An environmental improvement plan needs to be formulated for every city or major urban location. The magnitude and causes of pollution and the measures by which it can be mitigated also need to be understood. An unfortunate example can be seen in the case of Delhi. At a huge cost to the public, a major shift in fuel policy is being implemented via the use of compressed natural gas (CNG) for public transport. Several of my colleagues and I believe that what we see in Delhi is a mistake. At the end of the day, these expensive adventures barely scratch the surface of the problem. The switch to CNG is an extremely costly exercise. With LNG being imported and with the infrastructure needed requiring huge investment, only a very small part of the problem is being addressed at a very high cost. Thus, it is very important to carry out in a location such as Delhi modelling exercises, whereby the sources and receptors of air pollution, for instance, are clearly quantified and identified. This will allow sound decision making on what model would be the most appropriate with optimal allocation of resources for addressing the problem of poor air quality. Only by taking a comprehensive view of the problem can the most effective and suitable solutions emerge. In the case of Delhi for example, the importance of afforestation of the surrounding areas cannot be underestimated. Nor can you deny the need for reliable power supply. The number of diesel generator sets in the city is staggering. But in the city nobody bothers about them. This is the direct result of very poor power supply in the capital. Of course, we also need clean and affordable fuels.

4. Policies for Rural Prosperity

It is important for us to ensure that the government implements policies which bring about prosperity in rural areas. It is not enough to merely redistribute income and bring about an equitable distribution of wealth. It is also important to create opportunities. This is where new technologies can play a vital role – whether it is information technology, renewable energy, agricultural biotechnology or other methods by which forestry can be promoted. An appropriate mix of these technologies, decided on by local communities, can have a huge impact, not only in rural areas but also in urban centres. A state like West Bengal and its capital Kolkata, with their enlightened leadership and enormous potential, should be able to bring about technological improvements for the benefit of the poor. An exercise in at least four or five cities can be undertaken as models of institutional improvement and environmental governance.

I will cite an example which also has a personal connection. The Darjeeling Himalayan Railway Heritage Foundation spearheaded the campaign to get the Darjeeling Himalayan Railways declared a world heritage site by UNESCO. Through a variety of circumstances, I was made chairman of this foundation. We have carried out a very detailed study of the environment in the Darjeeling region. There are enormous opportunities for bringing about improvement. The city is grappling with several problems. A great deal of the sewage generated is being dumped into its water streams; solid waste is being littered along its roads and other locations; its forestry and bio-diversity are under threat. In my view, Darjeeling with its unique railway, can become one of the most favoured tourist locations. Effective eco-tourism can draw people from all over the world to this site. I hope major initiatives will be taken to improve Darjeeling's environment and transform it into a model location. I feel deeply concerned about doing something for this region.

I would like to end with one final observation. A major asset that Kolkata has, which is not often written about, is its social capital. References have been to the culture of the city. The fact is that Kolkata is a microcosm of global society. If its social capital can be combined with technology, financial capital and human capital, I think this city can take off. While you find several voices ready to criticize Kolkata, what they do not realize is the value of its social capital. Despite a large poor population, there is always harmony in living between the rich and poor. Outsiders are accepted with open arms in the city even when local residents feel deprived and underprivileged. This demonstrates the social cohesion of Kolkata and its people. Thus, there is reason to feel optimistic. In the new millennium, Kolkata promises to be a resurgent city. It can become a city that stands out. With all the initiatives that are being taken to improve the environment of the city, with the involvement of civil society and the local community, I have no doubt that Kolkata will become a shining example for the rest of the world.

4

Urban Environmental Management

Some Observations*

Asim Dasgupta

Dr George Mathew, my colleague Mr. Manab Mukerjee, very distinguished participants at this International Conference on Urban Environment Management, I am indeed thankful to you for giving me this opportunity to interact with you on this issue. May I congratulate you for convening this conference and extend my warm greetings to all the participants, particularly those from outside the state who have taken the trouble to come and attend. I hope your stay here has been pleasant and fruitful.

I will make several observations on the theme of the conference. The first observation I wish to make is that when we talk about managing the urban environment, we should be referring not simply to the physical environment but also to the social environment. In other words, along with the issues relating to air, noise and water pollution and the handling of sewage or solid waste, we should be addressing issues relating to the access to livelihood, employment, health, education and, most difficult, to land ownership and tenure security within urban areas. This concern has been reflected in many papers presented at the conference. Land ownership and tenure security are very important. Eventually, we have to address ourselves to the question: Whose city is this? Whose city are we talking about? Who are the people staying there? These aspects have to be viewed through the eyes of the common people in urban

---

* Valedictory Address

areas. Thus, both the social and physical aspects of the environment are of equal importance.

My second observation is related to the first one. That is, the management of the environment is linked to the density of the population. Whether it is provision of safe drinking water, handling of sewage or other facilities, the cost of each of these may increase exponentially with the density of the urban population. The cost of provisioning of services increases with the density of population. In the management of the urban environment, we should lay a proportionate emphasis on decentralization in the entire process of urban growth. Decentralization does not imply neglecting the environment of the metropolitan cities in India, Pakistan, Thailand or Sri Lanka. They too are very important. But effective management of the environment in metropolitan cities in turn requires decentralized growth of urban dwellings. By this I do not mean growth of four or five satellite towns, or of district headquarters or even subdivisional headquarters. Urban decentralization should mean paying adequate prior attention to the provisioning of urban facilities all the way down to block headquarters.

A block is a conglomeration of around 100 villages. These are the areas that require good urban planning, and its corollary, land use planning. Land use planning does not merely imply dispersal from metropolitan-to district- to block-level towns – potential towns. It also means that within metropolitan areas, we should be very careful about the increase in the density of population in big cities. For example, municipal sanction for high-rise buildings should be discussed very carefully. In fact, this conference may seriously consider the proposition that medium-level high-rise buildings be permitted, not very tall buildings. Moderate high-rise buildings, i.e., up to the fourth floor or thereabouts, are manageable. Buildings beyond this height create all sort of physical and social problems. These relate to the greater concentration of people, as also the protection of water bodies and open spaces in urban areas. Land reforms are needed in urban areas as well to protect the interests of the common people, those who live in slums and other underprivileged areas. When high-rise buildings are constructed, it is the relatively affluent who benefit and those displaced are the poorer people. It is almost a battle, excuse my saying so along class lines, as in the case of land reforms. This needs to be given serious thought, otherwise the city, in terms of ownership, will go out of hands of the common people. Then whose water supply will we be worrying about? People who can buy bottled water any time? What is the point of thinking about the provisioning of safe drinking water for them?

My next observation relates to the provision of safe drinking water. This is among the most important environmental challenges, in both rural and urban areas. A survey of the morbidity pattern in West Bengal shows that nearly 50 per cent of the diseases are caused by the lack of provisioning of safe drinking water. So I am glad that attention has been focused on this issue at this conference. It is not simply a question of providing or increasing or augmenting the total supply of water. It also has to be safe. In case of West Bengal, it has to be arsenic-free; not tainted by any fluoride poisoning. As far as possible, it should be surface-based. Less dependence on ground water is always helpful. Even if there is a good aquifer to begin with there are problems later on. In additions, water should be available in adequate quantities. Availability of safe drinking water in metropolitan areas ideally should be 40 gallons per capita per day. In other areas, it should be at least 25 gallons per capita per day. In Kolkata, the availability is now 44 gallons per capita per day. In Mumbai, it is about 40 gallons per capita per day, in Delhi 33 gallons and in Chennai 18 gallons. Provisioning of the total supply is only half the battle. Its distribution is as important, especially in areas where poorer people live, including slums. In the maintenance and provisioning of pipes, the local industry and labour can be fruitfully involved.

My fourth observation pertains to sewage and drainage. Lots have been said at this conference on how to handle solid waste, which is a very positive contribution. In this area, there is considerable potential for recycling. I will give the example of the outskirts of the city of Kolkata. When you are travelling from the airport to Park hotel, you might have seen greenery on both sides. Do you know what these tank waters really are? They are sewage-fed fisheries. Now this is a century-old tradition in Kolkata. Kolkata's sewage, and I am sure this is comparable to the situation elsewhere, goes through three major pumping stations: a central pumping station, the one at Palmer Bazar and the North Choubagha pumping station. The pumping stations gorge out the sewage into canals, which transport it to about 200 shallow ponds where the sewage water is stored. In three weeks' time with sunshine and blue-green algae there is a natural oxidation, rendering the stinking sewage water crystal clear! We got a molecular biological test done by Kolkata University and the water was absolutely safe. More importantly, these small ponds act as water receptacles. When there is a heavy downpour, they hold the water and act as buffers against floods. If there is excess water, these sewage-fed fisheries also have a channel to carry the water into adjacent areas. Because of the extremely fertilized water, half the fish supplies of our city are obtained from these fisheries. The process taps nascent oxygen, preserves environmental balance and is almost cost-free! Compared to this, a mechanical oxidation pond is extremely expensive and needs to be in operation all the time. The recycling route is almost free of cost with multiple benefits. It can be tried in any medium-sized town. In a very small town, water accumulation will not be adequate for the purpose. But in a city like Karachi, or any city in Sri Lanka or Thailand, with a gradient on one side, it is a very useful exercise.

This brings me to solid waste management. Dhaka's example has been adequately corroborated. There have been similar experiments in Kolkata, going back a hundred years. This explains the lush green vegetation on the outskirts of the city. Kolkata's solid waste is dumped there. It takes six months for the waste to be solidified and get transformed into absolutely first-rate organic manure. On these lands, the cropping intensity is high with 10-11 crops in a year. Since the land is over-fertilized, several crops need to be grown. Otherwise, the crops could get burnt. The city's good vegetable supply is thanks to the recycling of solid waste. By pushing the boundaries of constraints and transforming them into assets, the rate of growth is accelerated. There is very well-known growth model in economics developed by von Newman, who was the Head of the Department of Mathematics on Princeton University for several years. According to this model, a high growth rate of production in the economy can be achieved by converting constraints into assets. This is amply demonstrated by the urban sewage and solid waste examples cited above.

Dhaka's experience of handling solid waste is applicable to South-East Asia as a whole. Here, the issue is also one of management. An important question concerning solid waste management is: How can there be coordination between local governments, NGOs/community groups and private initiatives? The starting point of the entire exercise of solid waste management is collection from households. There also has to be collection at the ward level without contamination of the surrounding area. In Kolkata, an attempt has been made since the last eight years or so by the concerned councillor (if he or she is a good councillor and wishes to be reelected) to be present early in the morning when the handcarts arrive. People should contain their garbage in some form and place it in the handcarts. There should be the trucks waiting, to which the garbage from the handcarts can be transferred. The role of municipalities can vary. The absolutely localized aspects, can be handed over to NGOs while maintaining transparency and a competitive spirit. Municipalities can manage the subsequent stages.

I will now take up the question of roads. We often are very impressed by the major roads linking different parts of our big cities. They are usually kept in a good condition. But in urban areas it is also very important to pay attention to the state of the smaller inner networks, just like in the case of water supply. Roads going into slums and poorer neighbourhoods also need to be maintained properly. How can this be ensured? Who will take the responsibility? Can these be local involvement? These are very important questions. Another crucial area is traffic management. This issue has implications for the environment. It is not simply a matter of constructing 'x' number of flyovers and so forth. It is about evolving innovative ways of managing traffic through community participation. The entry of trucks into, and their passage through, cities needs to be regulated. Other than constructing truck terminals, norms need to be laid down about the use of roads by mass transit vehicles, private vehicles and commercial vehicles. I would like to give two examples of traffic management. During the Durga Puja festival, the 6 million population of Kolkata touches nearly 10 million. Handling such a huge population in a city where the road space is only 6 per cent of the total area is an enormous challenge. The car population over the last 10 years has more than doubled. Yet, during those festival days the traffic in Kolkata is manageable. It is because trucks are not allowed to enter. They are held outside the city. Only local small vehicles are permitted entry. Therefore, truck terminals should be constructed in a decentralized manner.

In Indian cities, signal lights at intersections are not adequate for managing traffic. People's participation is also important here. As you approach Kolkata from the airport, you find a high density of population in Rajarhat municipality. Here, the traffic management is not merely the responsibility of the police; the local youth also help out. They are paid on a contract basis for standing at junctions and monitoring the traffic. Their participation helped to streamline the traffic within just a week. They, in turn, get employment. This motivates people to get involved. In Kolkata 9 out of 10 vehicles on the city's roads are private vehicles, like passenger cars, and not fully occupied. Only 10 per cent of the traffic consists of public vehicles, carrying the bulk of the city's population. Thus, norms regarding use of road space need to be evolved. In the central business district, called BBD BAG, we once tried an experiment. There are three roads approaching the business district – one is only meant for private cars, the second is for mass transit vehicles and the third for mixed use. The green light was kept on longer for mass transit vehicles. On that day, there was no traffic jam in the central business district. Traffic jams are basically caused by cars, make no mistake about it, and of course the indisciplined drivers of buses. But they can be disciplined if traffic rules are strictly implemented. This needs to be given priority because during a traffic jam, toxic vehicular emissions shoot up, particularly in the central business district. According to studies conducted by the Environment Department of the Government, it is the most polluted part of Kolkata city. Without sounding romantic, we need to seriously think about the pedestrian plaza concept for the central business district. Here, even ministers will have to walk, along with industrialists and ordinary citizens.

It will simply mean one additional kilometer of walking for the richest in the city. This is approximately the distance the average commuter cover on foot anyway. The pedestrian plaza is a great equalizer besides bringing down levels of environmental pollution considerably. As you know, it has taken different forms in different countries.

While on the issue of air pollution, I would like to emphasize the importance of finding a balance between maintaining environmental standards and protecting employment. To the extent air pollution from industries can be contained without dislocating factories, it is less disruptive for the workers. The option of using pollution control devices to check industrial emissions must be thoroughly exhausted before dislocating plants. This will better serve the interests of social balance. A related area of concern is noise pollution which has not been discussed at this conference. It can be tackled by mounting an effective social campaign against it. A strong message needs to go out that under no circumstance will noise beyond permissible decibel levels be tolerated. The local community can play an important role during festivals.

Next, I would like to talk about the health dimension. Take the morbidity pattern. I will not go into the malnutrition aspect, which plays a very major role in morbidity. Looking at the environmental dimension, a little more than 50 per cent of the morbidity can be prevented by supplying safe drinking water. From the point of view of treatment, decentralized local bodies and municipalities will help by preventing the undue concentration of patients in metropolitan city hospitals. We conducted a survey of Kolkata's five major teaching hospitals. We found that 85 per cent of the patients in these hospitals shouldn't have been admitted there. They should have been treated at the block, district, or municipal hospitals.

I would urge those of you who have the time to visit a hospital in New Barrackpore which is close to the airport. There and in adjacent municipalities an experiment is being tried out. It is similar to the experiment conducted 15 years back with CUDP-III or Calcutta Urban Development Project. For every 200 households, there is a lady health worker appointed on contract with a monthly remuneration of only Rs 500. These women are trained in preventive health care like inoculation, and in treatment of minor ailments. In the case of major health problems they refer the patients to hospitals. At municipality-run hospitals, for patients above the poverty line, nothing is free. The charges are a little higher than those at the usual government hospitals but much less than the charges at private hospitals. The doctors and the paramedical staff get comparable salaries, but each of them is hired on a contract basis, a social contract. At the end of the years, there is an open meeting where the patients give feedback on whether the doctor and the staff were available. Contracts are renewed only in the case of those who perform satisfactorily. The New Barrackpore municipal hospital, which started with only a maternity ward and OPD, now has a range of health care facilities. A few nursing homes close by had to close down; at an adjacent government hospital, there are hardly any patients. In fact, this hospital attracts patients from adjacent districts. The infant mortality rate (IMR) in West Bengal is 50 per 1000. I think, this state ranks third at the all-India level; after 48 per 1000 in Maharashtra, and 15 per 1000 in Kerala. But the IMR for New Barrackpore municipality is just 9 per 1000. It is probably the lowest in the country. And maternal mortality is zero. Five other municipalities in the state have emulated it with comparable success. This is something you can see for yourself. Make a surprise visit. It is an example of the middle path – i.e., between the IMF line where patients have to pay for everything and the centralized over-subsidized government delivery system.

Another issue that I would like to take up is education. I think it plays an important role. We have experimented in this area much before Madhya Pradesh. The primary schools in panchayats are run by teaching assistants – we don't call them teachers since it can create legal problems. They are in the 40-plus age groups. They have the same qualifications as the regular teachers but are appointed on contract. It is cost-effective. The teaching assistants have to come and teach regularly for the contract to be renewed. Attendance is 100 per cent; the dropout rate is very low. They are doing very well. We have extended the scheme from the primary to the secondary level. This experiment can be tried in urban areas as well. We have taken a decision to extend its reach. Municipalities either be fully responsible for this programme or they can involve NGOs to assist them. At the secondary stage, it is not sufficient to merely impart general education. It must be combined with technical education and vocational education. This is very important from the point of view of self-employment and finding appropriate livelihoods. Employment generation should be an important part of urban planning.

I now return to the question of tenural security and slums. There have been quite a few papers and ideas on slums at this conference. I would like to mention that there are marginalized populations, not only in slums but in other fringe areas of the city as well, including the canal sides. When canals are being dug, how should these inhabitants be taken care of? Can they be optimally or temporarily relocated? These slums are extremely integrated. For slum development, the answer is not high-rise buildings for the so-called benefit of the poor. If concrete dwellings are built in slums, the underprivileged will be displaced by the middle class. No doubt about it! Our suggestion would be to make the slums a little more livable in terms of providing all amenities – water supply, roads, electricity, sewage, drainage, education and self-employment. Upgrade the dwelling place without replacing it by a four-storied structure. Such constructions will totally change the social milieu. The most human qualities are found in slums. Whenever collections are made for flood victims, those who come forward and empty their vessels are the people living in slums. Do not evict them from the city. Improve the slums in the manner I have suggested from my own experience.

Finally, I wish to make some observations on the monetary and managerial dimensions. What should be the role of the government, private initiatives and NGOs in the formulation of plans, provision of funds and implementation? A recurring theme in many papers presented in this conference is that municipalities are starved of funds. In West Bengal, we have tried to provide funds to municipalities out of the state exchequer. We have 123 urban local bodies. The total funds provided to them over the last three years is Rs. 8520 million. But one-third of the government funds have still not been utilized. As far as the availability of funds is concerned, if the government means business, it can always allocate funds. And if the municipalities are serious about taxing property and not taking a lenient view, they can generate funds. So the argument that state governments and municipalities are short of funds and therefore the need to bring in the private sector in general is not valid. However, there may be other grounds for tapping private funds. On an average a small municipality in our state should be getting at least Rs.10 million from the government. Even Rs. 20 million isn't adequate. Now I have an open mind on the involvement of the private sector. Municipalities and private enterprise can work hand in hand. But the partnership has to be within a transparent framework. This means that the municipality must hold regular elections. The process of advertising for bids must be an open one. Even the involvement of NGOs should be very transparent; their terms should be open to public scrutiny.

I'm emphasizing this for a very important reason. There are areas, such as slums, where the market sometimes fails or does not enter. In such areas, you need social action. It can take the form of municipal action or it can be NGO action; but it has to be properly coordinated. Because we need competition in social action as well. If there is transparency in the competition between municipalities, NGOs and private initiatives competition will better serve the overall social objectives. Keeping this in mind and the various points I discussed earlier, it is indeed possible to think in terms of good workable plan for municipalities.

The schemes can be of three types: first a trans-municipality scheme involving different municipalities or a municipality and rural areas. Many sewerage and drainage schemes belong to this category. Second, there are the trans-ward schemes. Most of the road schemes are of this type. The third type are the absolutely local, ward-level schemes. Whether it is water supply, roads, sewerage or drainage, this categorization is very important. Let's focus on the ward level. The ward committee is constitutionally mandatory. However, even in our state, where we try to keep a strong political vigil, I have to say that ward committees with very enthusiastic participation by the people cannot be found in all the municipalities. There are variations across municipalities. The councillor is very important because he or she represents the people. The ward committee too is very crucial. If the ward committee comprises people who belong to only certain sections of society, then you do not have wide representation. This is where political wisdom is required. You need to ensure a cross section of people on the ward committees. This is where I think NGOs can play a very important role at the level of ward committees. We have experimented with this in some cases.

First, there need to be open meetings. In case of clear weather, meetings must not be held indoors. If they are held outdoors, attendance is bound to increase. At these meetings, the funds available for the various sectors is specified and thrown open to discussion. There will be a lot of debate, even some bickering. I have attended committee meetings in rural as well as urban areas that have lasted for seven to eight hours without even electricity. People attend with lanterns, including the women, if the meeting is an animated one. In the interest of transparency, the views of NGOs, whatever they may be, should be placed at mass meetings – regarding the choice of schemes, technology and so forth. Once the ward-level schemes are selected - they include proposals for slightly bigger schemes; it may be for a road that does not end in a ward but connects with other wards – the people of the wards that stand to benefit have a right to propose them. So ward-level planning involves ward-level schemes plus proposals that get tied up at the municipality level. You have a mosaic that can be worked out in a 2x2 matrix for all the important sectors. Once implementation starts, there should be open discussions at the ward committee meetings about how much the municipality can do for itself? Can the municipality merely rely on its own workforce? Or does it need to get the work done on a contract basis? The municipality can invite bids from private contractors based on their performance. It can give the task to a contractual management. Or the municipality can delegate the job to NGOs. These options should be left open to the ward committee and the municipality. The experience will vary from one municipality to another, keeping in mind the ethos of competition and social cooperation. During implementation, a beneficiary committee must be constituted. The committee should include all potential beneficiaries, whether the implementing body is a municipality, an NGO or a private party. The beneficiary committee should submit a report – on the basis of which the final payment should be made. This I think is very important. And the accounts statement with the physical progress report should be placed in the public domain, whether the project has been implemented by a municipality, or an NGO or a private entity. Competition should be fair between elected bodies and NGOs and accountability should be there for all players. In the case of a municipality, the councillors are all elected. They have to face the people at the end of five years or so. Normally in our state, a councillor does not even earn Rs 1,000. We have allowed CAG (Comptroller and Auditor General) audits for municipalities. But an NGO is not an elected body. So where does its accountability lie? Now regular NGO staff would be earning more than a councillor in many cases. In our state, I am totally in favour of, I'm sure you will agree, CAG auditing of NGOs – their source of funds, how they are constituted. So competition should be among equals. Genuine transparent competition between municipalities, elected bodies, NGOs and private enterprises will ensure good performance. Such an approach can unleash a new force. It's not about municipality versus NGOs or private sector. Rather, it is participation of all in a transparent manner, with accountability to the people. As a finance minister, I would recommend CAG audit for all the players.

PART I

Theme Papers

5
Urban Environment Management
A Local Governance Perspective

Mohit Bhattacharya

Introduction

The National Commission on Urbanization (1988), India's last most comprehensive urban inquiry commission, referred to the degradation of urban environment in the following terms:" ... these urban centres have ... generated the most brutal and inhuman living conditions, with large sections of the citizens (almost half in Mumbai and Delhi) living in squatter settlements. The overcrowding in the slums and the desperate lack of water and sanitation leads not only to severe health problems but to the abject degradation of human life." Most "Third World" cities have been afflicted with problems of unregulated and insanitary squatter settlements, pollution of water and air, high-density living conditions, and the almost daily traffic jams caused by ever-rising numbers of motor vehicles jostling for space in the midst of competitive users. It is not the purpose of this paper to present a demographic profile of cities and towns. Nor will there be any attempt to depict the state of urban civic services in a comparative perspective. What this paper seeks to achieve, even if sketchily, is to relate the issue of urban environment management to the grass-roots institutional questions. Urban environment normatively belongs to the institutions of local governance. In reality, however, supralocal actors have virtually superseded local institutions and, in many instances, pre-empted their functional jurisdictions. It is the contention of this paper that the disjunction between urban environment management and local institutional governance is largely responsible for the continuing crisis in the management of the local environment. Broadly, this paper traces the historical evolution of environmental concerns in the global context, situates local government in the enterprise of environment management, and deals with the problematic of local capacity building to grapple with the complexities of urban environment management.

Two overarching concepts hold together the disparate discussion in this paper first, carrying capacity; second, urban governance. The former has been in wide use in the realms of urban planning and project formulation. The latter, which is often traced to World Bank's lending conditionality, subsumes a set of actors in collective problem solving, including both formal governmental agencies and non-governmental actors like NGOs and CBOs. Local governments' leading role in environment management has been clearly enunciated in the historic Agenda 21 of the Rio Declaration (1992). In India, the 74th Constitutional Amendment Act (CAA) of 1992 made it obligatory for urban local bodies to undertake micro-level planning, keeping in mind the interdependencies in local areas, their environmental concerns and the optimum level of local resource use.

Special mention should be made in this connection of the Global Campaign on Urban Governance launched by Habitat in 1999, with its emphasis on the "inclusive city" idea – involving everyone, regardless of wealth, gender, age, race or religion, in a participatory process of governance to face up to the challenges of urban poverty and environmental degradation

Carrying Capacity

"Carrying capacity" refers to the finite capacity or the limits of the natural environment, both as a reservoir of resources to support human consumption and as a sink to assimilate the residuals or wastes. Thus, carrying capacity-based planning needs to deal with the management of the throughput: that is, the size and nature of human activities leading to resource consumption and waste generation, as well as the supportive resource base and assimilative capacities of the environment. In other words, the planning approach may require optimization of human demands in relation to the manageable supply of environmental resources.

In recent times, the carrying capacity-based approach to urban planning has been accepted as a tool for sustainable development of human settlements. The need for the urgent formulation of policies and action programmes towards sustainable development of human settlements has been brought into sharp focus by Agenda 21 of UNCED (United Nations Conference on Environment and Development). However, India is yet to operationalize the Agenda at the field level. The very concept of carrying capacity-based planning is new in India, not to speak of the availability of tools and techniques to operationalize the concept at the level of urban and regional planning.

The urban environment has to have the capacities to assimilate, that is, to manage and recycle various wastes – air, water, land, and noise pollution – generated by its population and economic activities. It also requires capacities in terms of various natural and socio-economic resources and infrastructure – namely, land, water, energy, transport, social amenities, economic base and institutions – to support its population and economic activities. The two broad dimensions of environmental carrying capacities of urban areas may be categorized in terms of: (a) waste assimilative capacities; and (b) capacities in respect of various environmental resources and infrastructure.

Through a comparative analysis of case studies across seven developing countries of Asia, Europe and Latin America, Leitmann (1993) has been able to group the major urban environmental problems into thirteen functional areas, which can be linked specifically to the waste assimilative capacities of air, water, land, biological and cross-media components of the urban environment. Interestingly, noise pollution or the acoustic environment of cities has not been identified as a major environmental problem area. Resource classification is helpful for preparing an inventory of the existing resource base of an urban region and analyzing the impact thereon of development scenarios.

National Environmental Engineering Research Institute's (NEERI) classification of regional resources helps to identify the various environmental components in respect of which carrying capacity assessments are important for the sustainable development of cities. This classification covers a whole range of resources, such as air, water, land and biological resources;

economic resources for urban industrial activities; organizational/institutional resources for both economic development and urban amenities and infrastructure, industrial technological base and urban infrastructure and social amenities resources.

The Organization for Economic Cooperation and Development (OECD) identified a set of urban environmental concerns in 1978, along with their indicators. These include commercial, health, educational, recreational, transportation and protective services; employment, which has been considered as the key concern in respect of the urban economy; air, water and acoustic environments; urban solid and hazardous waste management; and land quality and urban landscape, which refers specifically to conservation, open spaces and landscape amenities. Although social and cultural concerns have also been listed, no specifications for this component have been articulated.

The listing of policy goals helps to identify different urban environmental components for planning and development purposes. The United Nations Centre for Human Settlements (UNCHS) has developed two sets of indicators: urban indicators and housing indicators. The urban indicators, organized into five different modules, focus on the environmental components of the urban economy – specifically, poverty, employment and productivity; life quality of the population in terms of life expectancy, infant mortality, literacy, health, education and social integration; urban utilities and services, namely, water, sewage, electricity, telephone, transportation infrastructure; and local government institutional resources. The housing indicators deal specifically with various policy goals for housing development.

The urban environmental indicators developed by Leitmann, as earlier mentioned, for the World Bank UMP Project on Rapid Urban Environmental Assessment are organized under modules covering the demographic characteristics of the population, including, growth rate, density, life expectancy, infant mortality and economic status; natural environment, including its biological components; land characteristics, in terms of drainage and topography, and climate; urban land use; air quality and energy use in terms of pollution emission, energy consumption and pollution control mechanisms; water resources and supply and waste generation and management.

Urbanization is a companion of development. However, cities in Africa, Asia, the Near East and Latin America have had to live with the challenges of congestion and pollution and with concentrated poverty and uncontrolled sprawl. These factors impede productive growth and, at worst, stifle it.

Cities in the developing world are expanding at a rate of 62 million inhabitants annually, which is equivalent to adding a country the size of France or Egypt every year. As many as one inhabitant in four now lives in poverty in these cities. *The estimated 400 million urban poor (as of 1996) represents an enormous challenge to the urban planner*. The challenge of providing them with critical infrastructure services joins the need to reverse environmental decay. Alongside this, there is the need to ensure social cohesion as one of the primary concerns of the action agenda. Safe, accessible drinking water is an all-too-common example of a basic commodity denied to the urban poor. Where cities fail to provide it, street vendors sell water as a priced commodity. The result is a heavy informal tax on the poor people.

Without basic services, slum dwellers in Accra, Quito or Surbaya are forced deeper into a marginal existence, unable to realize their own potential or to contribute their full worth to their communities. In India's main cities – Mumbai, Kolkata, and Delhi –between 40 to 50 per cent of the residents live in informal congested, often illegal, settlements and unhealthy slums unserved by city facilities. Of the eight million Manilans, an estimated one in five is a squatter – many of them living below the municipal poverty line. Providing for the needs of such numbers is a colossal task.

Community response

Local people's involvement in all stages of environmental planning and implementation has been crucial for attaining good results in many places. No one knows better than the residents of blighted neighbourhoods do which services they lack and what level of intervention will improve their health and well-being.

Indonesia, for instance, has had 25 years of experience with its Kampung Improvement Programme (KIP). Over the years, KIP has reached nearly 15 million people and spread to some 300 local government units around Indonesia. The programme focused on the provision of basic services; water, sanitation, shelter and roads. Strong partnerships and allocating roles and responsibilities to the local and central governments – while engaging the community to participate in and "own" the programme have been key to the success of the programme. Similar efforts have been undertaken in Bolivia, India, Jordan, Morocco, Pakistan and the Philippines, among other countries. In the Kolkata Metropolitan Area, a similar effort was successfully launched and completed in the third phase, of Calcutta Urban Development Programme- or CUDP III, in 1983-84.

Residents of Aashiana, a colony in Lucknow, proudly report that they are managing their own garbage collection, although it costs each family Rs 30 per month. They have also persuaded the Lucknow Development Authority to let them build a vermi-composting pit in the green belt. Govindpuri, a resettlement colony in Delhi, is well ahead of the posh colonies of South Delhi in its solid waste management plan. Residents there now segregate garbage; and green rickshaws collect organic waste, while red ones pick up the inorganic component. Their *yuva manch* (youth forum) performs street plays to educate people about the new system. Even Patna, famous for cowsheds and garbage mountains, has begun to change. Tired of waiting for the municipality, some middle class *mohallas,* with support from voluntary agencies, have privatized street cleaning and garbage collection.

The most impressive story of collective action is that of Civic Exnora in Tamil Nadu. A resident of Vadapalani Road in Chennai says: "Our street used to be one big garbage dump; the bin outside our home was always overflowing, as the corporation van often did not show up. My neighbour would set the garbage on fire, but the smoke irritated my asthma. I would douse it with water. So we fought all the time.

"One morning, the dustbin disappeared and a brightly-painted cart stood at my door with a boy in uniform and gloves. Called the "street beautifier", he taught us to separate our garbage at home. Each morning, he empties the organic waste into the green section of his cart and the recyclable waste into the red. Later, he takes the cart to our Zero Waste Centre, where he empties the organic waste into a storage tank for conversion into compost. He sells the recyclables and the compost to augment his income. I pay Rs 20 a month, and our street is now spotlessly clean. Where there used to be garbage outside each home, we have now planted trees."

This change came about thanks to the efforts of Exnora Club. Started by M B Nirmal, a bank manager, this civic movement is so successful that it is rapidly spreading across South India. It now covers 40 per cent of Chennai city and 75 per cent of its suburbs and has clubs across Tamil Nadu and in the three southern states. Exnora Club's 17,000 street chapters provide clean, scientific garbage collection services to approximately 1.7 million homes. Many clubs are solving other civic problems (sewage, street lighting, water supply) through their municipality. Exnora was recognized by the UNCHS in 1996 among the 100 Best Urban Practices (*The Times of India*, Kolkata, June 30, 2002).

Such programmes have to enlist many participants in both the planning and execution phases. Not only should central and local governments share the burden and involve the affected communities from the beginning, their projects need the active participation of local utilities and non-governmental organizations (NGOs). In service delivery and environment management (e.g., solid waste), the private sector has also been acknowledged as a major participant.

When the urban poor are helped to help themselves, the provision of basic services goes beyond bricks and mortar into the forging of a stronger community. Investments that are sound on economic grounds can also pay significant social dividends. Slum families affected by upgradation programmes in Manila between 1976 and 1985 had invested a large amount of money in improving homes which they felt were truly their own property.

Suggested Action Programme

To bring about improvement in the urban environment today, a few concrete actions should engage our attention. Lead in gasoline should be phased out at the earliest. As mentioned above, bringing basic services, above all clean water, to all city-dwellers should be the first task. Here, slum-dwellers need not be excluded; rather, they should be made partners. This measure is affordable. Many investments to reduce emissions of dust, soot and smoke from industry and power plants yield high returns. To proceed effectively with environmental improvements, a process of community participation is needed. Even measures such as removing lead from gasoline, which are relatively inexpensive, require determined public education efforts to win wide support. City officials have to join environmental action coalitions to mount awareness campaigns. Consultations must begin early to achieve community consensus on how to share the costs and what benefits to expect. Consultation that leads to visible progress and concrete results will build growing public support necessary to maintain or increase the momentum.

Urban environment management involves both networking and internal managerial upgrading. Resources must be managed more effectively by pricing urban services more accurately; building partnerships with the private sector to manage and finance urban infrastructure; and building much stronger local government institutions to operate the nuts and bolts of city management. It is the chronic disability of local institutions that has led to deeper environmental crisis and the virtual bypassing of these institutions by the creation of ad hoc agencies. This leads to further weakening of local institutions.

The Rio Earth Summit in 1992 called upon governments to mobilize their communities for the localization of Agenda 21. This UN Conference on Environment and Development (UNCED) went further. Its thematic guidance favoured community-led participation in the approach to environmental improvement, introducing thereby an emphasis on bottom-up participation rather than a reliance on top-down planning practice. This new environmental planning and management (EPM) has also been advocated in the Sustainable Cities Programme (SCP), organized as demonstration projects in designated cities by UNCHS and the UN Environment Programme (UNEP).

The analytical-conceptual framework for the localization of Agenda 21 and its companion piece, the Habitat Agenda from the UNCHS Habitat II Conference, Istanbul, 1996 – has its foundations in the idea of sustainability. Sustainability is, of course, the guiding principle of the UNCED Agenda 21. In terms of localizing the notion of sustainable urban development, the Agenda called on local governments to undertake a consultative process with their populations and to achieve consensus on Local Agenda 21 for their communities. The application of localized sustainability depended on the elaboration of schema for EPM. This means that local authorities, acting within broad-based community participation, will provide legal, institutional, organizational and action-oriented directions for environmental enhancement and protection. Along with other relevant precepts derived from the 1990s, there will be emphasis on social inclusion, social equity, capacity building, and the participation of major stakeholders in local governance.

Following the Rio meeting in 1992, UNCHS published its reflections on sustainability (UNCHS 1998). This publication sets out the nature and extent of environmental problems and their causes. Most significantly, it points out the need for improving urban governance and continuous reform in urban planning. In the following year, UNCHS took the rationale for the reform of urban planning a few steps further, UNCHS had also obtained legitimacy for sustainability in urban environments from its Habitat II Conference in Istanbul, 1996. Habitat II had an affinity with UNCED, with of course greater influence by the developing countries in setting the agenda and guiding policy. The UNCHS position statements were in terms of developing a symbiotic relationship between economic development and the environment. Bridging any identified resource gaps; steering policies toward the reduction of environmental poverty; and placing some reforms in governance within an enabling framework were the other major concerns.

Role of International Agencies

Since the United Nations Conference on the Human Environment, held at Stockholm in 1972, and Habitat II at Vancouver in 1976, various UN agencies and the World Bank have been making efforts to incorporate environmental considerations into their development activities. One of the major outcomes of such efforts is the publication of *Environmental Guidelines for Settlement Planning and Management* in three volumes designed to gather available knowledge about the relationship between the natural environment and the built environment; and to provide guidelines to planners and decision-makers to help them use this knowledge in settlement planning and management. The guidelines became necessary with the realization on the part of administrators, planners, managers and policy-makers responsible for human settlements in developing countries that safeguarding the natural environment was not an impediment to the development of settlements. Further, environmental management was perceived as a means of ensuring sustainable development. These instructional volumes offered guidelines for institutionalizing EPM and integrating other disciplines covering metropolitan planning and management, and regional planning and management.

According to SCP, environmental degradation is a common but not inevitable outcome of urban growth. It is caused primarily by inappropriate urban development policies and ineffective planning and management, as also due to inadequate consideration of the constraints and opportunities of the natural environment. In other words, "environment" as such need not be identified as the problem – the root cause lies in policy and operational lapses. To address these problems, the SCP laid down a number of principles and viewpoints, such as environment as a resource to be managed in a sustainable manner; striking a two-way relationship between development and environment; and the need for new types of political and managerial action to cope with the problems.

In tackling local conflicts, organizational limitations and environmental issues, the SCP follows a bottom-up approach, focusing on local solutions. To fulfil its aim of providing an improved EPM capacity to a city, the SCP follows requests from cities in need of assistance. A city is selected as a demonstration project only in response to a clear demand by the concerned city authorities. Then the SCP helps to formulate and develop a demonstration project, which takes on the character of that city's sustainable city project. Such a bottom-up approach leads to a strong sense of ownership of the project among the local partners, resulting in a sense of commitment by them to see the project through a successful conclusion. This is followed by the implementation of an EPM process through its designated phases, which begins with the preparation of the city's environmental profile. In this exercise, the main motivating force is the original impulse from local city authorities, who should first articulate their felt need for environmental improvement action. SCP is not to be imposed from outside – it should follow the innate urges of the local authority.

Local Government

Following this approach, in recent years, local government reforms have been undertaken in those countries around the world that are directly geared to institutional strengthening for the purpose of enhancing the coping ability of town and cities. Although a milestone in the better management of cities and towns, this approach has also significantly shifted the focus of power in urban governance and local political dynamics. Localization has gained a much wider meaning. It proposes new ways of doing things, including coordination among various government departments, between different levels of government, among different urban regional development agencies, promoting privatization, partnerships and the mobilization of local communities. Local governance is thus conceived in much broader multi-actor terms, which marks a distinct shift from conventional "municipal" or single-actor thinking.

As earlier stated, the use of the term "sustainable cities" and "sustainable human settlements" was much in evidence at

Habitat II; the second UN Conference on Human Settlements held in Istanbul in June 1996. Within the framework of "sustainable development", five broad categories of environmental action can be identified for the purpose of assessment of environmental performance of cities (Setterthwaite 1999).

1. Control infections and parasitic diseases and the health burden they place on city populations. This is popularly known as the "brown agenda" or the sanitary agenda. It includes adequate provision of water, sanitation, drainage and garbage collection for every one living in the city.
2. Put in place safeguards to tackle the chemical and physical hazards within the home, workplace and wider city.
3. Ensure a high-quality city environment for all city-dwellers – in terms of the amount and quality of open space and protection of the natural and cultural heritage.
4. Minimize the transfer of environmental costs to the inhabitants and ecosystem surrounding the city.
5. Ensure steady progress towards "sustainable consumption". That is, the goods and services required to meet everyone's consumption needs should be delivered without undermining the environmental capital of the nation and the world. This involves use of resources and consumption of goods imported into the city, and generation and disposal of waste by the city, in a manner that is compatible with the limits of natural capital and not by transferring environmental costs to other people.

Of these, the first three categories generally fit within the conventional mandate of local authorities. But the fourth and fifth categories are more problematic. They usually fall outside the jurisdiction of local authorities responsible for the environmental management of cities.

Intervention Modalities

The developing world has been searching for suitable intervention methods for poverty alleviation and authentic local development, including environment management, over the past four or five decades. Management strategies have been changing from time to time to suit the local conditions; new elements have been incorporated in the policy perspective. Through the fifties, sixties and seventies, strategies of community development, integrated local development and basic needs followed one after another in quick succession. "Participation" and "empowerment" as an intervention methodology evolved in the 1970s as a response to the inadequacies of previous top-down strategies. "Putting the last first", to quote Robert Chambers (1983), has been widely acclaimed as a desirable strategy.

The *Human Development Report* (UNDP 1993) commented in the same vein: "People's participation is becoming the central issue of our times." In this context, community participation as a means of mobilizing local popular energy, initiative and resources has occupied the centre stage, in so far as local problem solving is concerned. It is "part of a wider strategy to promote savings to target services only towards those who have been identified as being most desperately in need of them and to shift the burden of resource provision away from the public sector towards communities, including communities in greatest need themselves." And this is to be community-propelled, not external expert -propelled.

The *World Development Report 1999/2000* talks of new institutional responses that are needed in a globalizing world. Institutions are conceptualized as sets of rules governing the actions of individuals and organizations, the interaction between relevant parties and the negotiations among the participants. Institutions of good governance based on consensual, participatory and transparent processes are critical for development and should encompass partnerships among all elements of civil society. On the one hand, *globalization* tends to reflect the progressive integration of the world's economies; on the other, localization reflects the growing desire of people for a greater say in their government. Harnessing civil society, the local trust and efforts (social capital), has now been accepted as a basic requirement of authentic local planning and problem solving.

Municipal innovation

The locality movement has thus captivated planning attention in recent years. The last decade of the past century witnessed many examples of municipal institutional innovations to promote "sustainable development" – a concept we owe basically to the famous Brundtland Commission (World Commission on Environment and Development 1987). Long-term programmes for improving the environment and reducing resource use and waste were launched by many cities in Europe and North America as a sequel to the Rio Conference on Environment and Development in 1992. As stated earlier, this conference produced the landmark document titled Agenda 21. The term is also used to describe the sustainable development plans formulated for particular cities and localities. Local authorities in many cities have committed themselves to sustainable development goals and have been eager to share their knowledge and experience with other city authorities. In some Latin American cities as well as those in other regions of the world, city authorities have found innovative ways of addressing environmental problems. There is also a world-wide "healthy city" movement in which local authorities in more than a 1,000 cities (e.g., Chittagong in Bangladesh) have sought new ways to interact with different actors and interests for the promotion of health and prevention of disease at the locality level. Within the Habitat II held in Istanbul in 1996, the World Health Organization *Report on Creating Healthy Cities in the 21st Century* was specially prepared for a focused dialogue on the need for a healthy urban future. Briefly stated, the Healthy Cities Programme addressed two aspects of governance at the local level:

1. Technical aspects, involving local resource mobilization, plan formulation, technology application and resource allocation
2. Representational and participatory aspects, including participation, channels for popular representation, and increased transparency and accountability in the functioning of local authorities

In respect of need assessment and priority setting, the Healthy Cities project has (*a*) considered the views of the communities involved, in terms of their perceived needs and priorities, and (*b*) made technical assessments based on available health statistics and known epidemiological linkages between environmental and social conditions.

To cite an instance, Chittagong, the second largest city in Bangladesh with around 2 million people, is currently running a Healthy City project for which there is a steering committee chaired by the Mayor. The committee includes representatives from different sectors and groups. A project office has been created within the City Corporation. In addition, there are zonal task forces or specific plans and actions in particular areas; and sectoral task forces responsible for specific plans and actions in housing, water, sanitation, etc.

Columbia, which was one of the participants at the Rio Earth Summit of 1992, encouraged the framing of "bio-plans" for the country's municipalities in terms of the guidelines of Agenda 21. For instance, Manizales municipality became internationally known as a local authority that had successfully prepared a bio-plan with the following goals in mind:

1. Stimulate knowledge of the local and regional eco-system.
2. Conserve the municipality's natural and cultural resources.
3. Increase energy efficiency in the running of the city.
4. Improve urban living conditions with special focus on environmental sanitation, integrated waste management and recycling, and improved security for citizens in public spaces.
5. Apply appropriate technologies in industrial production
6. Provide for the efficient supply of public services and transport
7. Ensure environmental education for popular participation.

An elaborate monitoring and evaluation mechanism was devised to strengthen the efficacy of the municipal environmental action plan. For each of the eleven communes (territorial units of the city), plans have been prepared and implemented. Ten database indicators have been developed, which reflect overall environmental quality. Each of these indicators is assigned one of three grades: Red (R) indicates problems; Yellow (Y) warns about possible problems; and Green (G) suggests good quality.

As shown in Box.1, this is an interesting device, which allows a visual representation of where environmental problems are concentrated in terms of sectors and areas within the municipal jurisdiction. The system is known as the environmental quality traffic light (E & TL). Admittedly, it requires considerable sophistication with regard to data utilization, collection, processing and information. Manizales municipality has experienced many difficulties in acquiring, processing and analyzing different kinds of data relevant to the formulation of E & TL. Nevertheless, it is a very imaginative method for assessing local environmental management.

Box. 1: Environmental Traffic Lights: How each commune scored on composite indicators of environmental quality

Composite Indicators	1	2	3	4	5	6	7	8	9	10	11
	(Different areas/localities)										
Social well-being (including indicators of health, education, social security and income)	G	R	Y	Y	R	Y	Y	G	Y	Y	Y
Quality and accessibility of public services (based on provision of piped water, sanitation, electricity, gas, public telephones)	G	Y	G	G	Y	G	G	G	Y	Y	G
Housing quality (based on quality of construction, density and provision of community services)	G	R	Y	G	Y	Y	Y	G	Y	Y	Y
Health environment (based on air and water quality and extent of noise and pollution)	G	Y	G	G	Y	Y	Y	G	G	Y	G

Possibility of enjoying public space (based on, among other things, access to parks and ecological reserves)	G	R	R	Y	R	Y	G	G	Y	Y	Y
Aesthetic and symbolic value of landscape (related to richness and variety of natural and built environment)	G	R	Y	Y	Y	Y	Y	Y	Y	Y	Y
Physical security of the area (based on level of risk from earthquakes, eruptions, landslides and floods)	Y	R	Y	Y	R	Y	Y	Y	Y	Y	Y
Citizen security (based on frequency of assaults, murders, traffic accidents, vandalism of public space)	Y	R	Y	Y	Y	Y	Y	Y	Y	Y	Y
Quality and efficiency of transport	G	Y	Y	Y	Y	Y	Y	Y	Y	Y	Y
Citizen participation in environmental issues (related to extent of participation in different projects and programmes)	R	G	Y	G	G	Y	G	G	Y	Y	R

R: Red, which indicates problems; Y: Yellow, which gives warning of possible problems; G: Green, which indicates good quality.
Source: Setterthwaite (1999).

Indian Scenario

India has been a late starter in responding to either Agenda 21 or other international recommendations. The 74th CAA may be construed as a vital provision mandating the strong role of local governments in micro-level planning, which should include "environment" as a major item.

Recently the Good Urban Governance Campaign (India Launch) was jointly organized by the Ministry of Urban Development and Poverty Alleviation and the United Nations

Charter for Human Settlements in New Delhi (4-6 September 2001).

The campaign focused on six key areas: (*i*) decentralization, (*ii*) municipal finance, (*iii*) urban environment, (*iv*) intergration of the poor and the marginalized, (*v*) transparency and civic engagement, and (*vi*) municipal management capacity building.

Based on the experience of cities, a National Plan of Action for taking the Urban Governance Campaign forward has been developed for implementation over a period of three years. The Government of India will continue to take the lead in coordination, guidance and advice. The 74th CAA must be reviewed, especially Article 242(W) regarding the functional domain of municipalities, which should be implemented in letter and spirit. Among the other major recommendations in this regard are:

1. Voters should have the power to recall councillors if they fail to deliver.
2. Electoral reforms are needed at the level of local bodies. Members of Parliament and Members of Legislative Assemblies should not be included in the local bodies.
3. The employees of local bodies should be accountable to the elected leadership.
4. Local bodies should frame a Citizen's Charter, indicating standards of service. The local bodies should introduce Good Governance Report Cards for effective service delivery. Citizens should have the right to information on all public matters related to local bodies.
5. A comprehensive city information system should be developed, which can be effectively accessed by citizens.
6. District planning committees and ward committees should be given enough financial powers.
7. The decentralization efforts of the 35 metro cities, down to the ward committees, should be reviewed.
8. Municipal capacity building should cover (besides municipal strengthening) political, administrative, and grass-roots level functionaries, including NGOs, CBOs, the private sector etc.

Strong policy-level commitment at higher levels is required to translate these recommendations into action, specially through adequate capacity-building at the local municipal level.

In the Indian context, environmental management needs to be prioritized with special reference to the metro cities and the Class I cities, which together bear the brunt of urbanization. It is in these large towns and cities that more focused attention has to be paid to the "environment", both in terms of maintenance management and futuristic planned development. Maintenance management of solid waste, water supply, drainage and sewage, to cite a few services, would go a long way towards creating a cleaner urban environment. A large-scale slum within a city has to be treated as a specialized area management problem. This approach has paid rich dividends in Kolkata Metropolitan Area under CUDP III. Since micro-level urban planning is now a constitutional responsibility, a definitive bottom-up approach to municipality planning is now widely discussed. Environmental concerns need to be woven into a holistic area-level plan of a municipality. This will necessitate considerable technical strengthening of urban local bodies and networking among many agencies, such as development authorities, CBOs and NGOs.

REFERENCES

Annez Patricia 1998, "Livable Cities for the 21st Century", *Society*, May/June.

Chambers Robert 1983, *Rural Development: Putting the Last First*, Longman.

Leitmann Joseph 1993, *Rapid Urban Environmental Assessment: Lessons from Cities in the Developing World,* vols. 1 and 2, Urban Management Programme, World Bank, Washington DC.

Mcgill R 1998, "Urban Management in Developing Countries", *Cities*, vol. 15, No.6.

National Commission of Urbanization (NCU) 1988, *Report of the National Commission of Urbanisation, Vol. I,* NCU, Government of India.

Pugh C (ed.) 1996, *Sustainability: The Environment and Urbanization*, Earthscan, London.

Setterthwaite David, 1999, *Sustainable Cities*, Earthscan, London.

United Nations Development Programme (UNDP) 1993, *Human Development Report,* UNDP.

World Bank, 1999-2000, *World Development Report,* World Bank, Washington.

6

Poverty and Environmental Sustainability in Asian Cities

Issues, Policy Perspectives and New Initiatives

Amitabh Kundu

1. Introduction

Issues concerning sustainability of human settlements have come to the centre stage of policy debates in developing countries, at least since the nineteen eighties. A distinct change can be noted in the perspective of several international as well as national development agencies that have shifted their focus to environment related issues. This high-profile shift occurred at a time when the world community was passionately debating the issues of global warming, ecological degradation, etc., that sought to set limits on economic growth for the sake of survival of humanity. Many development planners working in the urban field had become apprehensive that the limited budgets available for developing infrastructure in towns and cities will be hijacked by non-urban projects and that the emphasis will shift from development to *macroenvironmental* issues. Luckily, several international agencies, including the World Bank, came forward to underline the importance of the microenvironment for the sake of sustainability of development. This led to an emphasis on the *brown agenda*, which called for larger investment in urban infrastructure such as water supply, sanitation, etc., particularly in slums and other low-income colonies. The agenda, in a way, has countered the demands of the global environment lobby and stressed the need for allocating substantial resources for tackling microenvironmental problems.

Unfortunately, not much serious thought has gone into understanding the meaning of sustainable development and how to achieve it in relation to the ongoing process of urbanization in the countries of Asia. Most of the environmental programmes launched through international initiatives have put forward a standard package of solutions. These have been introduced without paying adequate attention to regional socio-economic specificities. Restructuring the system of governance, particularly the legislative and administrative system, has often been perceived as the most crucial element in the solution package for meeting the challenge of urban crisis and environmental problems. It is argued that the governments in most Asian countries have opted for some kind of planning framework, which has hindered the functioning of the market, particularly for the provision of civic amenities. This has created barriers to prices being determined by the logic of the market and dampened investments in infrastructure and amenities, resulting in the precipitation of urban crises.[1]

The need to free the market for microenvironmental services from the regulative framework of the state, facilitate the private and joint sectors to invest in these services and enable local governments to take development decisions has been voiced in many Asian countries since the early eighties. A consequent shift in policy perspective can be seen in terms of the reduction in public sector intervention, the marginalization of the physical aspects of master planning, and the selection of capital-intensive projects at the city level based on the possibility of resource mobilization from private entrepreneurs or capital markets. The simplification of the legislative system and flexibility are also being advocated as significant components of the package of "technical solutions" (Harris and Fabricius 1996 and Gunatilake and Perera 1994). Remedies are being sought also in terms of efficient and transparent management of urban projects.

Many of the international agencies have taken the position that the existing legal and institutional structures cannot take up the responsibility of linking the cities to the "borderless" world (Ohmae 1991). These structures have no "capacity to deal with the demands and pressures that are being generated as a result of opening up of the economies" and that the system of urban management has remained insensitive to the needs of globalization. Cities are thus being seen as the products of complex forces, some generated from within the countries but mainly those exerted from the external world over which individual cities have little control. The message that comes across from this new perspective is that the responsibility of local leadership should not be limited only to the provision of basic amenities but should also include taking an *aggressive posture* to attract private capital, both from within and outside the country (Stren 1994). This, it is argued, cannot be done without drastically modifying the complicated legal system inherited from colonial or feudal regimes in Asia (Lee 1994 and Sivaramakrishnan and Green 1986). Economic liberalization and political decentralization are thus seen as harbingers of sustainable settlement structures, which, in turn, are expected to result in overall regional development.

The implicit assumption of this framework is that governments in the countries of Asia are in control of the urban development process but somehow the laws and regulatory mechanisms are not yet working very well. Its policy implication understandably is that all these countries should be overseen by governments with uniform "legal structures and administrative organs" (OECD 1994 and World Bank 1998). This perspective has been responsible for these governments not giving much attention towards the creation of local structures of accountability. The process of selection and implementation of urban projects has thus remained insensitive to economic and environmental considerations at the micro level. Virtually no effort has gone into understanding the underlying processes and forces that are creating cities in the less developed world and, hence, discovering what might be truly effective in guiding development for the benefit of the people. Consequently, the wider conditions of the world have come to undermine local priorities.

Serious doubts have, however, been raised about the efficacy of this new policy package to meet the widening deficiencies in infrastructure and basic amenities for all sections of the population. More specifically, it is apprehended that the new perspective may endanger the sustainability of development in many of the Asian countries (UNRISD 1995 and Kundu 2000). Unfortunately, research studies focusing on the impact of urban structures on environmental parameters or those linking the nature of governance to sustainability are hopelessly inadequate for developing countries. What is perhaps more worrying is that in the absence of such data, "the environmental problems that tend to get highlighted are those that are easily measured and those that affect middle and upper income groups" (Hardoy, Mitlin and Satterthwaite 1992).

It is further apprehended that in ensuring *global responsibility*, city governments often get tied down with conditions or structures, forcing them to underplay microenvironmental problems (Kundu 1994). They are increasingly facing a conflict between serving the needs of the general population, particularly of the poor and making investment in specialized services required by more sophisticated international business interests. In the absence of appropriate information systems and institutional backup, vested interests have been able to capture the political process, resulting in marginalization of the interests of the poor. There are, thus, ample reasons to be skeptical about the capacity of municipal governments in Asia to meet the needs of the microenvironment in urban areas while responding to the demands of globalization and decentralization (Kundu 2000, Coit 2000).

Importantly, worsening global debt patterns in the developing countries and nationally and internationally imposed austerity measures have dictated deep cuts in both new and recurrent spending on urban infrastructure. With major public sector investments by local agencies being effectively ruled out and the maintenance of existing infrastructure being neglected, the overall urban decline has become more or less endemic. The physical duality of extreme poverty and extreme wealth is, therefore, likely to become even more evident in the indebted globalized cities of Asia (PADCO 1990).

It is, thus, evident that the basis for the optimism in the new package of solutions has not been examined empirically. Given the contradictory claims made by researchers, the present paper is an overview of the major development initiatives launched in the past two decades in Asian countries, which have a bearing on the functioning of urban local bodies and examines their implications. An attempt has been made in section 2, which follows the present introductory section, to analyze the efficacy and implications of the new system of maintaining municipal budgets and the devolution of funds from state to local-level agencies. The third section examines the implications of the proposed institutional restructuring for enabling municipal bodies to undertake the responsibility for development planning. The fourth section discusses the initiatives of assigning contracts to private agencies for the provision and maintenance of basic services. The limited success of slum networking projects, implemented within the framework of community participation, has been examined in the following section. The sixth section analyzes the innovative methods adopted at the local level for mobilizing resources from the capital market and development-cum-banking institutions and the conditionalities imposed by lending agencies and financial intermediaries. The seventh section is focused on India as a case study. Here, the availability of select amenities like water supply, toilets and electricity (which are not explicitly incorporated in the official definition of poverty in India) has been analyzed at the city and town levels in relation to the trend and pattern of urbanization and the recent trends and policies on decentralization and infrastructure financing. In the last section, an attempt has been made to summarize the results and conclusions, examining the validity of the new perspective of governance in ensuring environmental sustainability.

2. New System of Maintaining Municipal Budgets and Devolution of State Funds Based on Performance

In several Asian countries, the federal governments until recently have been supporting the local bodies in their budgetary expenditures through a gap filling approach. This approach has been criticized for making the local bodies lax in matters of resource mobilization and for promoting inefficiency (Mathur 1994). Also, the system of providing funds by the state for developing urban infrastructure through government departments did not distinguish between a town and a city based on their tax raising efforts or efficiency. In the era of globalization, when the development strategy should be guided by the principles of efficiency and return on investment, this policy is gradually being abandoned or modified, wherever in existence.

The new approach is to promote infrastructural and industrial investment in cities and towns that can pay for themselves. The idea is to create an institutional and legal structure that would enable the *most deserving cities* to have access to resources. The need for increasing transparency in the functioning of local bodies, particularly in the accounting of their earnings and expenditure has, thus been emphasized. Further, transfer of funds, sanctioning of loans from national and international organizations, mobilization of resources through issuance of bonds, etc., are sought to be linked to the performance of the bodies, which is to be measured in an *objective manner*.

Unfortunately, there are serious conceptual and operational problems in building performance indicators using the budgetary figures of municipal bodies and other local agencies. The concepts and categories used in the budgets and their coverage vary widely across regions within a country and even across urban centres within a region. Identical ratios obtaining for different cities and towns, or for the same town over time, do not imply the same situation. This is due to differences in the system of budgeting.

Importantly, the capital earnings and expenditures of the municipal bodies are dependent on the delegation of functions by the central or state\regional governments. The variations in the levels of capital expenditure are, therefore, determined, at least partly, by the distribution of functions between the state and local governments. As a consequence, the inter-municipal comparison of the sectoral distribution of capital investment is not very meaningful.[2] The comparison of current earnings and expenditures, too, is fraught with problems. Administrative expenses for different sectors, for example, are often posted under the head of general administration, although the practice of showing these under the concerned sectors is also not uncommon. A large number of such anomalies of serious magnitude can be noted through micro-level investigations in almost in all the countries of Asia.

Accrual of surplus or deficit for an urban local body (ULB) in any given year can thus at least partially be attributed to the *internal dynamics* of the local body or an *official requirement* to show less expenditure in certain functional areas. What is, therefore, urgently required is standardization of the definitions and working out of the financial indicators in an unambiguous manner for the purpose of temporal and cross-sectional comparison at least within a country.[3] Unfortunately, national-level data generation agencies in most countries have made no serious attempt to standardize the municipal budgets for bringing about comparability of financial data across time and space.

Interestingly, a large number of financial agencies have come up on the Asian urban scene that have made clear-cut prescriptions with regards to the budgetary system of the municipalities. They want the local bodies to segregate the earnings and expenditures for different sectors and maintain separate accounts. This has become mandatory if the latter want to mobilize funds from international sources or the capital market.[4] This, along with the demand for transparency, is likely to put a halt to the system of cross-subsidization of *economically unattractive* sectors or of poor areas in the city. It would also restrain the local bodies, from undertaking capital or current expenditures of certain types out of their own revenues.

3. Assigning Planning Responsibilities to Local Bodies

Research studies in many of the Asian countries suggest that local bodies are simply not equipped to take up the responsibility of planning, and especially the launching of capital-intensive projects. Considerable expertise is required to identify the infrastructural and industrial projects appropriate for the growth of a city or town, assess their environmental implications and mobilize resources for these. Given their difficult financial situation, it is unlikely that these bodies will be able to strengthen their planning departments by recruiting technical and professional personnel in the immediate future. Assistance for this purpose is unlikely to come from the central or state governments since they, too, are facing serious financial problems and are unable to augment their professional staff (Siddiqui 1992, Nurul Amin1989 and Adhikari 1994).

The only choice for local bodies would then be to resort to financial intermediaries and credit rating agencies. A number of these agencies have come up in recent years in the private sector with assistance from certain international organizations. The major considerations in the preparation of projects by these agencies are the concerns of the companies that would provide the finance, commitments by the borrowing agency to ensure cost recovery, legal and administrative restructuring for making the agreement work, etc. Understandably, the "investments which are clearly linked to enhanced revenues"[5] are being preferred. The criteria canvassed by the agencies thus impose a definite financial perspective and discipline on the local bodies that may be at variance from that needed for the integrated development of the city or regional economy.

Assigning the responsibility of planning to local bodies thus does not automatically enable them to prepare projects based on a development perspective for the town or region. Preparing a master plan to answer the needs and aspirations of the local people in different social and income brackets and then breaking it down to a set of meaningful projects is a different exercise from that of identifying projects with high credit ratings that would attract corporate investment. In an era wherein obtaining funds for investment from the private sector or the capital market has become the critical factor, it is not surprising that the exercises of preparing *grandiose Master Plans* has been thoroughly discredited.[6] Given this emerging scenario in the nineties, the option of going for project preparation in formal or informal consultation with interested companies or stakeholders, through the intermediation of the financial institutions, seems to be an easy way out. Local governments in large cities have found this option hard to resist.

3. Management Contracts for Services to be provided by Private Agencies

Deficiencies in urban infrastructure and basic amenities have acquired alarming proportions in Asian countries due to the declining trend in investment and financial support from state governments for local bodies. With the limited potential of these bodies for raising funds from their own sources, the gap between demand and supply of basic services has widened in recent years. This has forced them to work out alternative institutional arrangements, resulting in the emergence of private sector participation in the management of municipal services (Mathur 1999a).

Different forms of participation have been designed with varying levels of responsibility and cost sharing between private and public agencies. The most commonly used form is contracting out management of one or a set of services by the municipal bodies to private companies. It has been argued that this will result in substantial cost saving on the part of local authorities, at the same time, make it possible for the companies to conduct their business with profit. There are apprehensions, however, that such sub-contracting of services will lead to the dilution of social responsibility and the exclusion of the poor and vulnerable sections from the formal delivery system.

In order to ensure that the contractual agreements with private or joint sector companies are executed, supervised or monitored under strict regulatory control, new regulatory/ supervisory arrangements are being proposed. It may be pointed out that under the old system of state control, the public agencies had failed to meet the social objectives and cater to the needs of the poor. Designing and implementing regulatory controls in a liberalized regime will, therefore, be far more challenging, in case we are serious about these objectives. Importantly, UNDP and UN-HABITAT have launched the Urban Management Programme (UMP) with offices in many developing countries. The purpose is to create an enabling environment for such participatory management practices and assist local bodies in drawing up terms and conditions of the contracts, involving community participation in management. The programme also aids the local bodies to establish contacts with private companies and often conducts negotiations on behalf of the former.

It is difficult to empirically determine the benefits and costs of privatization of management, as it is a recent phenomenon. Reliable data on the revenue and expenditure of local bodies before and after privatization or profitability of the private agencies is yet not available. The critical information in this context will be with regard to the quality of services provided and the coverage of vulnerable sections of the city population, as noted above. Unfortunately, there exists no database covering these aspects. This will have to be generated through field surveys at the micro-level. It is very surprising that none of the studies conducted in connection with the privatization of services or subcontracting arrangements in Asian cities has focused on this critical issue.

The overview studies on the contractual agreements, their progress reports and other official documents reveal that there have been problems even in operationalizing the working arrangements and norms for the day-to-day functioning. This is understandable, because many of the countries have little experience in partnership arrangements. Several cities have got trapped in serious difficulties after the formal launching of the projects.

Solid waste management is one service where the local bodies have shown great interest in involving the private sector. Until recently, ULBs in most Asian countries had no mandate to launch capital projects for sewerage and garbage disposal. Now that the states are devolving these responsibilities, local bodies have sought to involve private companies through contractual arrangements. Besides, many citizens' groups and non-governmental organizations (NGOs) have shown a willingness to spare time or money for improving the system of garbage disposal. Success stories have come from a number of cities. However, the fact that a number of companies have taken up the responsibility for the provision of this service and are making profits or that the local bodies have reduced their financial burden through contracting is by itself no indication of the success of the arrangement. It would be extremely important to monitor the quality of the services; and, more importantly, the coverage of the urban poor, which sometimes has gone down after the contracting arrangement was introduced. Further, there has been a demand from the companies to hike the user charges commensurate with inflation, which often had to be conceded. Understandably, several low-income and slum colonies have resisted such moves, resulting in their exclusion from the delivery system.

In fact, it is a big challenge to design an institutional and legal structure that can oblige the company to maintain the same level of commitment to the poor as that of the municipal body. In the eventuality of being excluded, people in the localities have not been able to raise their voices or complain against the municipal body as the latter would take the plea of passing on the responsibility to an outside agency. This has already led to the fragmentation of responsibility and the service delivery system, with marginalized areas\ people remaining underserviced or unserviced. This may pose a serious health risk for the entire urban community and may result in the outbreak of epidemics, to a much greater extent than what has happened in the past.

Given the financial scenario at local level, the keenness on the part of the ULBs to subcontract solid waste disposal is easily understood. It would, however, be difficult for them to motivate private companies to take up this job without their paying a substantial compensation. Unfortunately, the ULBs are in dire financial straits and the very motive behind contracting out is to save on capital as well as current expenses. Consequently, many among the ULBs have negotiated with the private agencies and agreed to compensate them not in cash but in kind. In many cases, they have leased out land to the companies on a long-term basis. The companies, too, find this arrangement remunerative because they can thereby use the precious urban land for other purposes, earning massive profits. Many companies have taken up the job of waste disposal solely for the purpose of getting a hold on urban land for alternative uses, without going through the complex land acquisition procedures or paying market rates.

City authorities generally possess no technical competence or motivation to assess the cost implications of the arrangements with private agencies for the people or the urban economy in the long term. There has been no effort on the part of state governments to provide an information base or technical assistance to ULBs for assessing the long-term implications of land deals and helping the latter to enforce appropriate administrative safeguards. There exists no institutional network to which an ULB can link up for seeking guidance on technical matters like the technology proposed by a private company for disposing or composting garbage. On the other hand, voices have been raised against such possible institutional interventions on the pretext that this would undermine or impinge on the authority of local bodies.

In a liberalized environment and with the withdrawal of the central and state governments in local affairs, it is understandable that ULBs are required to shoulder a large number of social responsibilities. It is, however, unfortunate that despite formal pronouncements by the central and state governments from time to time and some financial support given to ULBs for these purposes, the city administrations have failed to reach the poor and vulnerable in any satisfactory manner. Often the municipal corporations or parastatal agencies have admitted to providing, directly or through private arrangements, differential levels of services to different areas within the cities or failing to cover marginal areas. Importantly, the better-off localities with higher tax collections are demanding high-quality facilities and even pressing for the creation of separate local bodies. This has made cross-subsidization in the provision of amenities within the cities extremely difficult in recent years.

4. Involving Community in Infrastructure Development and Provision of Services

Community participation in urban development projects has a history dating back to the fifties or sixties in many of the Asian countries. The projects in the initial years mostly covered a few slums in select cities and even here, these were not designed to cover the total urban population. Consequently, the solutions pursued were local in nature. Social mobilization was attempted by creating community groups and getting them involved in the implementation of projects at the grass-roots level. The community was thus viewed merely as an agent, providing support to state-sponsored development activities for their better implementation in the initial experiments of community participation.

In subsequent years, it was realized that community involvement not only results in effective implementation and a substantial reduction in operational costs but also leads to better designing of the projects. Consequently, the community was often involved not merely in project implementation and supervision of work but also in the designing stage. All these initiatives notwithstanding, community participation remained a state-sponsored activity in most countries until the late seventies.

Community participation as an instrument of development strategy gained currency again in the eighties as a consequence of the poor performance of public agencies and a growing deficiency in the level of basic amenities. This was widely welcomed as it was argued that the community could help not only in social mobilization but also in raising financial resources[7] that the local authorities very badly need. In many cases, it became possible to substantially reduce in project costs as the prospective beneficiaries provided their labour free or at a wage rate lower than the prevailing market rate. Further, the pressure from peer groups under the participatory arrangement resulted in better monitoring, more productive engagement of the beneficiaries of the project and better recovery of development loans, sanctioned to individuals. The community was mobilized not merely for making contributions in terms of ideas and labour but also for sharing a part of the capital and current expenditure. Only, there was demand for making credit available outside the formal institutional structure, which was beyond the access of the slum communities. In several cases, attempts were made to build mechanisms at the community level to ensure the timely repayment of these loans.

Many were critical of the participatory approaches mentioned above as these left the poor to themselves for designing and implementing the project or they were associated only with credit access and the loan recovery mechanism. It was further argued that when a facility is provided to a slum through a separate subsystem designed by the local community, independent of that for the whole city, it inevitably turns out to be of lower quality and consequently faces problems in the long run. As a consequence, community participation now is being brought in various innovative ways in urban development projects that do not minimize the role of professional skills or support from the state.

The most innovative form of participation, which has been hailed as a major achievement in the nineties, is the neighbourhood and slum improvement schemes through *networking* – launched by local bodies, with some support from the state or central government. The people in the project area are associated with all the stages in the development of the project and are given the overall responsibility of monitoring it as also maintaining the services. The basic objective of such networking projects is to make high quality techno-engineering expertise available to the people, within the framework of community participation and upgrade the slums in a manner that they get integrated "into the fabric of city life."[8] These involve linking of the subsystem in the slum with the city-level network, which promises to make a positive impact on the macroenvironment of the city. This approach, understandably, helps in mainstreaming the poor by underlining the fact that slum development is a critical component of overall urban development.

The networking projects are generally based on partnerships between the municipality, non-government agencies, community groups and the private corporate sector. These are mostly for the development of roads, pavements, water supply, sewerage, drainage and waste disposal facilities, street lighting, landscaping, etc. within slums. Besides, the projects envisage the involvement of community-based groups that play an important role in resource mobilization as also in organizing other social activities. Thrust is given to community activities by making NGOs partners in the project, right from the initial stage; they have a role in maintaining the facilities created under the project.

This new approach has been projected not as a purely welfare project but one that can be financially viable if the city administration can motivate the private sector to join in. In a few cases, private industries have been associated, which have eliminated or reduced the need for external financial support. The basic logic is that private industries will always have an interest in the overall hygiene and health of the city because that determines their business prospects in the long run.[9] They can, therefore, be persuaded to share the cost of slum improvement. This perspective has allowed the advocates of this approach to place slum-linked activities within the mainstream of urban development strategy.

There are, however, several issues and questions with regard to this approach that remain unanswered at an analytical plane. First, there does not seem to be adequate evidence that the projects based on this approach are replicable on a large scale. It is, therefore difficult to place it as an alternative to the slum improvement strategy in the country. Criticisms have also been levelled due to neglect or the low-key manner in which the involvement of the community has been sought or obtained.[10]

Critics have pointed out that in most projects, there is no serious effort to create community-level organizations and involve them in plan preparation and implementation. In the absence of strong community participation and private sector backing, the projects are often implemented by local bodies with the help of a few NGOs. As a result of the low-key operation, the concept of integrating all the slums into the basic fabric of city life has become a dream.

5. Raising Resources from Capital Markets and Financial Institutions

In the new era of globalization, most of the local bodies in Asian countries are finding it difficult to undertake any major investment project by borrowing money directly from banks or other financial institutions due to their weak economic position. They have problems in issuing *general bonds*. The capital market may not attach much value to the words of trust or assurance given by ULBs or development authorities due to their messy and precarious budgetary situation. It is also unfortunate that despite the euphoria surrounding privatization, the development projects proposed by these bodies are often perceived as non-viable by the private corporate sector. As a consequence, many of them have opted for resource mobilization using innovative credit instruments, backed by financial intermediaries (Mathur 1999b).

Unfortunately, ULBs have not been able to mobilize resources through *revenue bonds* – the major source of capital financing in United States and Western Europe. Under this system, repayment obligations are linked to the revenues generated through the projects, which are launched using the debt amount. However, many of the urban development projects do not yield the required rate of return, at least in the short run. The *general obligation bonds*, on the other hand, require a commitment of total budgetary support from the local bodies. Even this may not be of much help because of the same reason that the overall budgetary position of most of ULBs is not very strong. *Structured debt obligations* (SDOs), therefore, emerge as interesting arrangements through which bonds are issued on the condition that the borrowing agency pledges or escrows certain buoyant sources of revenue like octroi, property tax, etc., for debt servicing. This allows the repayment obligations to be given priority and kept independent of the overall financial position of the borrowing agency. SDOs are designed in such a manner that the repayment of loans is done through the committed revenue sources, irrespective of the overall budget, which may be in deficit.

Given the enormous deficiency in infrastructure and the need to mobilize resources, many of the local bodies in Asian countries find escrowing their specific sources of earning not adequate for meeting the debt service obligations. As a consequence, they are often required to pledge other sources of revenue and grants as well, which is done in the case of general obligation bonds. The bond holders, thus, enjoy the security of certain dedicated revenue streams and, in addition, are given the faith and credit pledge of the city government, thereby making the issue an SDO as well as a general obligation bond.

The issuing authorities are often required to adhere to certain additional conditionalities – like maintaining an average debt service credit ratio, pledging municipal assets with certain asset coverage ratio, creating a sinking fund, etc. – particularly if they are seeking a high rating from credit rating agencies. In case of any possibility of changes in the macropolicies that may affect the budgetary position of the concerned authority (like abolition of octroi), assurances of compensatory payment by the federal government have been stipulated. All these are precautionary measures to protect the interests of the investors.

Several international agencies operating in the urban sector have stressed the need to keep the issuance of bonds outside the regulatory framework of the government and have pleaded for the development of an independent capital market. Accordingly, a large number of ULBs are making efforts to get a favourable credit rating for the purpose of lifting funds from the capital market without the backing of the government. For this, they are striving to achieve a budgetary surplus by cutting down, among other things, *un-remunerative social expenditures*. The support of the federal government for financing capital projects has not been forthcoming easily in recent years. There have been enormous delays in getting its blessings in any form, which in turn, have increased the cost of issuing the bonds and of the project. To counter this, attempts are being made to design a *comprehensive debt limitation policy.* This will ensure compliance of the ULBs with certain norms. It will also set a ceiling on their borrowing capacity, based on their economic strength and repayment capacity, without the state having to give approval in every case. Market borrowing on the part of local authorities can thus be made *independent of the state government.*

Unfortunately, as a consequence of stiff competition among the ULBs as also political and civil pressures on them to restore the level of social expenditures, many may default on debt repayment. It is indeed very frightening to imagine how a local body would tide over a financial crisis if a project launched through bond money fails financially and the state refuses to bail it out, which it can do legally.

Another apprehension about developing urban infrastructure by raising funds from the capital market is the accentuation of inequality across regions as well as size class of urban centres. Understandably, a few among the large cities, with a growing economy and buoyant revenue sources, will be able to tap the market, bypassing all bureaucratic hassles, under the new policy. The smaller townships, however, will be in no position to mobilize resources, with the discontinuance of the practice of state guarantee. Consequently, the greater dependence on the capital market in future will increase intercity/town disparity in terms of per capita expenditure and consequently the level of amenities (Kundu et. al. 1999).

It is evident from the above overview that recent developments in the capital market while liberating local agencies from the control of the central and state governments, will place the former under more stringent conditions imposed by the trustees, commercial banks, etc., and governed largely by pure market logic.[11] Financial power will shift from the state governments to financial institutions, international donors and credit rating agencies, which through various innovative and complicated arrangements, will dictate the expenditure pattern.[12] They will also be in a position to determine the type of projects to be undertaken and, in certain cases, even the management system of the municipalities. It will, therefore, be extremely important to monitor the problems faced by many of the Asian cities in handling their budgetary earnings and expenditures after accepting the terms and conditions for raising resources from the capital market.

After the initial euphoria about infrastructure development through innovative credit instruments, the present thinking is for greater stringency and caution in assessing the debt repayment capacity of projects. Municipal legislation often does not incorporate a clear ceiling on local borrowings and even when it does, the limit is not adhered to uniformly. This has created apprehensions that unorganized local borrowing may lead to an undue rise in public debt. The limitations of local accounting and auditing practices have led to the demand for making the "disclosure requirements for municipal bonds more stringent than for other issues"[13]. This will result in the imposition of stricter financial discipline on the functionaries in the capital market. One should, therefore, expect greater commercialization and pure profit motives governing municipal investments in the coming years[14].

6. Changing System of Governance and Resource Mobilisation for the Provision of Basic Amenities in India: Implications for Urbanization

It is important to analyze the impact of the new system of urban governance and the initiatives for resource mobilization, as discussed above, on the level of basic amenities and the trends and pattern of urbanization, taking India as a case study. Importantly, the rate of capital formation for infrastructure has been extremely low in India, creating serious deficiencies in the availability of infrastructure in towns and cities over the past few decades (Kundu and Maitra 1999). The uneven growth of the urban population across states and the size category of settlements have further aggravated the problem. The shortages are particularly serious in urban centres with a population less than fifty thousand. It will, therefore, be interesting to see how the new perspective has affected or will affect the deficiency in basic amenities and the pattern of urbanization.

Significant variations exist in the percentage of households covered by amenities like electricity, drinking water and toilets, across size classes of urban centres. The average figures for urban India are low in the case of toilet facilities and electricity connection. In 1981, these were 58 per cent and 62 per cent, respectively (Table 1), but slightly higher for drinking water, viz. 75 per cent. The percentage had gone up to 64, 76 and 81, respectively in 1991(Table 2).

The variation across the size class of urban centres shows regularity and a distinct pattern. The percentage of households covered by each of the three amenities increases systematically with the size class of urban centres, except for class VI towns (Tables 1 and 2).[15] It is indeed very disturbing that in class V towns, with a population between 5,000 and 10,000, the percentages for households not covered by toilet, electricity and drinking water facilities were as high as 61, 41 and 28, respectively, in 1991 (Table 2). The corresponding figures for cities with more than 0.5 million population are just 22, 17 and 14 per cent, respectively. Similarly, while 14 per cent of the people in class V towns have none of these three amenities, the figure for the half million plus cities is less than 3 per cent. The percentage for households having all the three amenities in class III, IV and V towns is about 30 while that for class I cities is twice as large.

Table 1: Percentage of Households with Access to Different Civic Amenities in Different Size Classes of Urban Centres in 1981

Urban Centres	Electricity	Drinking Water	Toilet
Class I (1 lakh plus)	61.90	75.36	58.00
Class Ia (1 to 3 lakh	52.48	71.34	46.98
Class Ib (3 to 5 lakh)	49.29	61.46	44.87
Class Ic (5 lakh plus)	45.97	73.21	62.80
Class II	58.19	70.33	51.90
Class III	54.76	65.37	45.42
Class IV	48.95	61.75	40.23
Class V	45.74	56.43	37.93
Class VI	51.26	56.62	39.63
Urban India	61.90	75.36	58.00

Source: Household Tables, Part viii, A & B (i & ii) (for different states), *Census of India*, 1981.

There has been a significant improvement in the availability of basic amenities across the size class of urban centres during 1981-91. The increase, however, is much greater in class I cities compared to the towns in other size categories, as may be seen in Tables 1 and 2. The percentage of households with electricity, for example, has gone up from 62 to 81 in these cities. Similarly, the figures have improved from 75 to 84 in the case of drinking water and from 58 to 74 in the case of toilet facilities. The corresponding increases for other size classes of urban centres are much less. Importantly, in many of the less developed states, such as Kerala, Madhya Pradesh, Bihar and Orissa, the percentage of households covered by toilet facilities in towns with a less than 50,000 population has remained stable or registered a decline. Further, in Orissa, a drastic fall in the figure is noted even in towns with a population between 50,000 and 100,000 (Kundu, Bagchi and Kundu 1999). These certainly are disturbing symptoms of a structural malady.

Table 2: Percentage of Households with Access to Different Civic Amenities in Different Size Classes of Urban Centres in 1991

Class	Electricity	Drinking water	Toilet	All Amenities	No Amenities
Class I	80.69	84.36	73.67	59.45	3.45
Class Ia	74.84	81.00	63.08	49.01	5.31
Class Ib	78.01	77.28	63.92	49.13	5.97
Class Ic	82.57	86.14	77.69	63.51	2.65
Class II	73.50	79.17	57.58	44.16	6.26
Class III	65.69	75.45	49.56	35.20	9.12
Class IV	62.19	73.84	43.08	30.38	11.47
Class V	59.19	72.18	38.75	27.82	13.91
Class VI	69.72	78.59	47.44	37.52	9.31
Urban India	76.51	81.84	66.43	52.46	5.17

Source: Unpublished Household Tables (available in computer floppies) from Population Census (1991)

It can be argued that a greater incidence of secondary and high-valued tertiary activities in large cities affords their inhabitants higher levels of income. In small and medium towns, income levels of the people tend to be low due to a poor economic base, the lack of employment opportunities in the organized sector, a weak manufacturing sector and a high percentage of the workforce dependent on agriculture. This may also be inferred also from the fact that these towns have less than 15 per cent of their workers engaged in non-household manufacturing as against the figure of over 25 per cent for the cities with populations in excess of hundred thousand as per the 1991 Census. Though the tertiary activities in these small and medium towns are high, they generally have a low level of productivity. Instability in their economies is reflected in high fluctuations in their demographic growth over time. Most of the small and medium towns are not in a position to generate funds to provide civic services to all the sections of the population and stabilize their economic base.

The new system of governance, therefore, is likely to increase the disparity in the levels of services and economic infrastructure across states and the size class of urban centres. This will adversely affect the level of basic services in small and medium towns in backward states and their capacity to attract new economic activities or absorb future growth of the population.

Given the perspective of urban governance and the decline in the rate of improvement in urban amenities, particularly in backward states and smaller towns, it is not surprising that the annual growth rate of the urban population in the country has gone down from 3.1 per cent during 1981-91 to 2.7 per cent during 1991-2001. This reinforces the declining trend of urbanization in the country – the growth rate had decelerated even during the eighties compared to the all-time high growth of 3.8 per cent during 1971-81.[16]

More important than the decline in the growth rate of the urban population is the change in the regional pattern of growth (Table 3), as the nineties made a significant departure from the earlier decades. Developed states like Tamil Nadu, Punjab, Haryana, Maharashtra and Gujarat registered urban growth above the national average. West Bengal is the only exception whose growth rate was not particularly impressive. The backward states, on the other hand, have experienced growth either below that of the country or, at the most, at par with the national rate.

Urbanization has, thus, become concentrated in developed states. Even here, the larger cities have recorded relatively higher growth. This may, at least partly, be attributed to decentralization measures, whereby the responsibilities for resource mobilization and launching infrastructure projects have been given to the local bodies, as noted above. Large municipal bodies that have a strong economic base, particularly those located in developed states, have an edge, which has clearly been manifested in their high economic and demographic growth.

A strong lobby has emerged, particularly in large cities, pleading for the disbanding of all zoning restrictions, building laws and bye-laws so as to make the cities relatively independent of state and central-level controls. It is stipulated that decisions regarding location of industries, change in land use, etc., should be taken expeditiously at the local level. Empowerment of local governments and decentralization of development planning responsibilities, ushered in through the 74th Constitutional Amendment 1992, have given a boost to this lobby. These cities have a relatively strong economic base and, consequently, high tax and non-tax revenue-generating capacity, which helped them to attract more investment. As a result, the urban structure became more top heavy in the nineties. Indeed, only a few of these large cities and their immediate hinterland have absorbed a large proportion of the migrants coming from rural areas. The high growth of the top 35 metro cities, much higher than that of an average urban centre or class I city, confirms this proposition. The new perspective of urban governance and the growing dependence on the capital market has meant considerable concentration of industrial activity in and around a few large cities of the country. National and multinational agencies have also shown an interest in developing these cities as global centres.

Thus, inequalities in urban-industrial growth across the country are likely to become very sharp in the coming years, as development activities get confined to a few big cities and their hinterland. Small and medium towns, many exhibiting significant growth potential in the past, are unlikely to attract private investment from within or outside the country, as they lack infrastructure and basic amenities. The deficiency in the coverage of households by drinking water, toilets and electricity is unlikely to be reduced in these towns, as they would find it difficult to finance their development projects through internal resources or borrowings from the capital market. The fiscal discipline imposed by the government, credit rating agencies and financial

Table 3: Trends and Pattern of Urbanization Across States in India

India/State / Union territory	Urban Population 2001(Per cent)	Growth Rate	
		1981-91	1991-2001
INDIA	**27.8**	**3.1**	**2.7**
Andhra Pradesh	27.1	3.6	1.4
Arunachal Pradesh	20.4	9.9	7.0
Assam	12.7	3.3	3.1
Bihar (undivided)	13.4	2.7	2.6
Chandigarh*	89.8	3.1	3.4
Delhi*	93.0	3.8	4.1
Goa	49.8	3.9	3.3
Gujarat	37.3	2.9	2.8
Haryana	29.0	3.6	4.1
Himachal Pradesh	9.8	3.2	2.8
Jammu & Kashmir	24.9	3.8	3.1
Karnataka	34.0	2.5	2.5
Kerala	26.0	4.8	0.7
Madhya Pradesh (undivided)	25.0	3.7	2.8
Maharashtra	42.4	3.3	2.9
Manipur	23.9	3.0	1.2
Meghalaya	19.6	3.1	3.1
Mizoram	49.5	9.5	3.3
Nagaland	17.7	5.5	5.3
Orissa	15.0	3.1	2.6
Punjab	34.0	2.5	3.2
Rajasthan	23.4	3.3	2.7
Sikkim	11.1	-3.2	4.8
Tamil Nadu	43.9	1.8	3.6
Tripura	17.0	6.2	2.5
Uttar Pradesh (undivided)	20.8	3.3	2.8
West Bengal	28.0	2.6	1.8

Source: Census of India, 2001.

intermediaries, will make it impossible for them to undertake major infrastructural investments. Understandably, this problem will assume more serious proportions in the less developed states. With governmental investment in infrastructure and basic amenities declining over the years, the disparity within the urban economy is bound to increase.

Only a few large cities with a strong economic base will be able to raise resources through bonds and other innovative credit instruments and strive to emerge as global centres. It will, however, be erroneous to hold that these cities, whose average conditions are much better than that of smaller towns, will have no problems of poverty or deficiency of basic amenities. Importantly, the general health situation depends not so much on the average expenditure on water supply and environmental sanitation as that made in marginal areas or slums. In fact, the slum situation is critical for the overall health of cities. Congested living conditions in the slums and little improvement over the years can act as a flashpoint for the outbreak of epidemics. This may, therefore, become a major bottleneck in attracting industrial investment in the near future. Asian cities facing a threat of epidemics can hardly attract business houses or grow into global cities.

8. Conclusions and Generalizations

The overview of the new system of urban governance as also the innovative methods of financing projects for providing basic amenities in Asian cities reveals that given the resource crunch in the economy, projects have come to depend on capital market borrowing, privatization, partnership arrangements and community participation. Planners and policy makers have, in recent years, made a strong case for parastatal agencies as also the local governments increasingly relying on their internal resources and institutional finance with the objective *of* bringing in efficiency and accountability in their functioning. The analyses of the foregoing sections suggest that this changed perspective and the consequent decline in public investment are likely to accentuate the disparity in the levels of amenities across the regions and size class of urban settlements.

Furthermore, the poor are likely to get priced out of the various schemes that are being launched within the private and joint sectors or through subcontracting arrangements. The companies participating in the schemes will strive to bring about a change in the development perspective and force a dilution of the social objectives. As a consequence, the better-off colonies will get more attention as also a larger share of the investible resources due to their higher purchasing power and greater political clout. Public sector schemes will be made to depend on institutional borrowings and the capital market, often generating funds at the market rate of interest. All this will deepen the gap in the level of basic amenities between the rich and poor localities within cities.

Efforts are underway to develop the capital market so that only those parastatal and local-level agencies that are more efficient can mobilize resources by issuing SDOs and other credit instruments. An analysis of the arrangements worked out by financial intermediaries, including the credit rating institutions, for tapping the capital market reveals that these have severely restricted the functioning of ULBs. In several cases, they have been forced to pledge their regular earnings from octroi, grants from the state, etc., as guarantees for debt servicing. This has come in the way of ULBs fulfilling their normal civic obligations.

Importantly, the projects that are likely to be financed through the credit market will, by their very logic, be commercially viable and ensure profitability to the investors and other stakeholders. However, in the eventuality of projects failing to generate the desired rate of profit, the concerned local body may be left with no funds even for general administration since much of its revenue earnings will be diverted for paying the investors or for completing the project. It thus appears that the policy of liberating the local governments from the regulatory and legislative controls of the state will place the former under the direct control of financial institutions. This will be very desirable from the viewpoint of developing a few global cities in Asia, providing space for business houses, office complexes and dwelling units for the executive and entrepreneurial classes but may not answer the needs of basic amenities for the majority of the urban poor.

Unfortunately, much of the subsidized amenities provided through governmental programmes during the sixties, seventies and early eighties were targeted a few large cities and benefited mostly the high and middle-income colonies. There is no way such subsidies can continue in the more liberalized regime of the twenty-first century. However, withdrawing government support and relegating the provision of services to market forces and subcontracting arrangements will accentuate the inequality across regions and size class of urban settlements. Discontinuing subsidies may also increase intraurban disparity and lead to segmentation, particularly in metropolises and other large cities, with serious consequences for the microenvironment and the health sector. This, in turn, will discourage the private corporate sector, both within and outside the country, from setting up their base in and around these cities. Consequently, the entire process of liberalization and creation of a few global cities in the country may get jeopardized. Welfare arguments apart, even the efficiency arguments for launching institutional innovations can be disputed by taking a slightly long-term perspective on urban development.

Posing urban and rural poverty as dichotomous categories or pitting one against the other would be erroneous. And, so would be the issue of small versus large urban centres. Within a regional setting, a large part of urban poverty, particularly in small and medium towns, can be seen as an outflow of rural poverty. Also, many of these towns, located on the periphery of large cities, are absorbing the poor who are pushed out of the central areas of the latter. A case can, therefore, be made for launching programmes for the provision of employment and basic amenities to the urban poor, particularly, in small and medium towns, as a part of a strategy for regional development.

Inequities in the provision of basic services across states and size class of urban centres in India are likely to get accentuated as a consequence of the programmes of structural adjustment. A case can, therefore, be made for providing special assistance to the less developed states, which are not in a position to allocate requisite funds to their towns and cities for this purpose. In particular, small and medium towns in backward states need to be supported financially by the federal government for undertaking capital projects, as their economic bases are not strong enough to mobilize adequate resources. This would imply increasing the resources allocated for urban development, as recommended in the report of the National Commission on Urbanization (1988). There must, however, be stipulations to ensure that most of the funds towards the provision of basic services for the urban poor and slum improvement.

Decentralization of financial powers through legal and administrative initiatives is not sufficient for generating resources at the town level. These must be backed up by the actual devolution of powers and responsibilities and their use by municipal bodies. The management capacities of these bodies need to be strengthened by hiring more technical personnel and training the existing staff so that they can organize their affairs better, including the mobilization of tax and non-tax resources. Cross-country studies reveal that manufacturing activities at the town level show a strong relationship with the availability of infrastructure and services. One may, therefore, argue that the provision of amenities in small urban settlements, besides being a goal in itself, will help in diversifying their economic base and generating non- agricultural employment.

The current programmes for providing basic amenities envisage the involvement of local bodies, private agencies, non-governmental organizations and communities. Government funds constitute only a part of the total resources; the remaining is expected to come from institutional sources, private entrepreneurs and the residents of the concerned town or locality. This automatically creates a system wherein the agencies or people providing substantial funds enjoy a greater authority in designing the details of the project, its exact location, etc. The organizations providing loans, for example, will have an interest in recovering their money within a short period. Towards this end, they can modify the project in a manner such that its original objectives get undermined or sidelined. Within this framework of decentralization, the urban poor, who at best can contribute, their labour and a small amount, will have very little say in working out the details of the projects or their implementation. It is, therefore, important that a structure for plan implementation be created so that the private agencies financing the projects on basic services are obliged to function within the framework of a development plan. Also, there should be proper monitoring of schemes so that vested interests at the local level do not corner a large part of the benefits based on their financial contribution to the project.

NOTES

1 *India Infrastructure Report* (Government of India 1996) attributes the deficiencies in infrastructure and basic amenities in India to a low rate of investment in the private sector due to inoptimal market conditions and a restrictive policy environment.

2 A study by National Institute of Public Finance and Policy (NIPFP 1995), covering 293 municipalities in seven states of India, reveals that even the larger corporations do not maintain well-defined capital accounts and record a number of capital expenditures in their revenue account.

3 Some difference in the scope and coverage of categories across municipalities and corporations may, however, be inevitable because of their varying functional responsibilities and powers of taxation.

4 The Indo-US Financial Institutions Reform and Expansion (FIRE) project in India (see Mehta and Satyanarayana 1996), for example, proposes a budgeting discipline, requiring local bodies to have separate accounts for not merely capital and current expenditures but also for different sets of sectors. It advocates pooling of select sectors into "cost centres" and ensuring profitability at the level of these cost centres. Proposals have been put forward for imposing or enhancing user charges and restricting intersectoral transfers, such that many among the service sectors can become independent of the general municipal budget. These, it is argued, would go a long way in attracting capital investments.

5 Indo-US FIRE Project (1998)

6 Based on a survey of select ULBs in the states of Gujarat, Maharashtra, Rajasthan and Madhaya Pradesh, Joshi (1996) argues that "most municipal bodies lack the much-needed long-term planning — while small municipal bodies can not afford it, unfortunately it is absent in big municipal corporations. The survey of 19 municipal corporations indicates that 73 per cent, i.e., 14 municipal corporations, had no formal long-term planning process. This indirectly points to an even more dismal picture in small municipal corporations and municipal bodies."

7 "It has been demonstrated in India that, firstly, large-scale renewal can be undertaken using slums as opportunities (Indore) and secondly, huge resources can be generated from the communities and supplemented by local authorities, industries and NGOs (pilot projects in Baroda and Ahmedabad)." Verma K, Slum Networking Project (mimeo), Office of the Municipal Commissioner, Ahmedabad.

8 See Municipal Corporation of Ahmedabad (1998)

9 Chairman of the Indian desk at Price Waterhouse has, in the context of networking projects, observed that the concept of corporate social responsibility is actually based on "hard-nosed commercial sense."

10 Op. cit.

11 The agencies active on the capital market admit that "such stringent requirements may place some burden on the development of a municipal bond system Indo-US FIRE Project (1998)."

12 Madhaya Pradesh government, for example, has admitted to its ability to provide escrow cover to only a few of the proposed power projects although a large number of applications are lying with them (*Economic Times*, Aug 1998).

13 Indo-US FIRE Project (1998).

14 Indo-US FIRE Project (1998).

15 This is because many of these towns have been established or are managed by government departments, public agencies, military establishments or private industries. They, therefore, enjoy a special status and a higher level of amenities.

16 The trend goes against not only the popular perception of "urban explosion" but also questions the projections made by expert groups set up by various government departments during the eighties and nineties, besides those of the *India Infrastructure Report* and the Tenth Five Year Plan

REFERENCES

Adhikari, S. P. (1994): "Urban Governance in Nepal", A Background Paper presented at the Subregional Asian Meeting of Global Urban Research Initiative, Lonavala

Coit K 2000: "Government, Sanitation and Environmental Heath", UNRISD-N-AERUS Workshop on Cities of the South: Sustainable for Whom? Geneva.

Expert Group on the Commercialisation of Infrastructure Projects 1996: *The India Infrastructure Report: Policy Imperatives for Growth and Welfare*, Ministry of Finance, Government of India, New Delhi

Gunatilake, G. and Perera, M. (1994): "Urban Poverty in Sri Lanka: Critical Issues and Policy Measures, *Asian Development Review,* vol. 12 no 1

Hardoy J E, D Mitlin, and D Satterthwaite, 1992: *Environmental Problems in Third World Cities,* Earthscan Publications, London.

Harris N and I Fabricius, (eds.) 1996: *Cities and Structural Adjustment,* University College of London, London.

Indo-US Financial Institutions Reform and Expansion Project 1998: "Emerging Issues in India's Municipal Bond System: The need for Regulatory Control", *Project Notes*, No 10

--------- 2000: *Lessons Learned: Ahmedabad Municipal Bond and Water Supply and Sewerage Project*, USAID, New Delhi

Joshi, R. 1996: "Financial Management in Municipalities", *Urban India,* vol. 16, No2

Kundu A 1994: *In the Name of the Urban Poor,* Sage Publications, New Delhi.

Kundu A 2000: "Urban Development, Infrastructure Financing and Emerging System of Governance in India: A Perspective", *Management of Social Transformation,* Discussion Paper No. 48, UNESCO, Paris.

Kundu A, S Bagchi, and D Kundu, 1999: "Regional Distribution of Infrastructure and Basic Amenities in Urban India", *Economic and Political Weekly*, 34(28).

Kundu A, and S Maitra 1999, "Access of Urban Poor to Basic Services – An Analysis in the Changing Perspective of Urban Governance in India" in Mathur, O. P (ed.), *The Challenge of Urban Governance*, National Institute of Public Finance and Policy, New Delhi.

Lee, M. 1994: "The Cost of Second Best: The Price of Land Regulation in Asia", *Working Paper*, no 62, Development Planning Unit, London

Mathur, O. P. 1999a: "Governing Cities: Facing up to the Challenge of Poverty and Globalisation", in

Mathur, O. P. (ed.) *India: The Challenge of Urban Governance*, National Institute of Public Finance and Policy, New Delhi

Mathur, O. P. 1999b: "Fiscal Innovations and Urban Governance" in Mathur, O. P. (ed.) *India: the Challenge of Urban Governance*, National Institute of Public Finance and Policy, New Delhi

Mathur, O. P. (1994): "Responding to the Urban Challenge: A Research Agenda for Nepal and India", in Stern, R. (ed.) *Research in the Developing World*, vol. I

Mehta, M and V. Satyanarayana, 1996: *Pricing and Cost Recovery for Urban Services*, Community Consulting International, New Delhi

Municipal Corporation of Ahmedabad 1998: *Slum Networking Project* (mimeo), Municipal Corporation of Ahmedabad, Ahmedabad

National Commission on Urbanization 1988: *Report of the National Commission on Urbanization*, Vol. I, Government of India, New Delhi.

National Institute of Public Finance and Policy 1995: *Redefining State-Municipal Fiscal Relations*, vol. I (mimeo), NIPFP, New Delhi

Nurul Amin, A. T. M., (1989): "Macro Perspectives on the Growth of the Informal Sector in Selected Asian Cities", *Paper presented at the Asian Employment Programme of the International Labour Organisation*, New Delhi

OECD (1994): *Cities for the Twenty First Century*, Organisation for Economic Cooperation and Development, Paris

Ohmae, K. (1991): *The Borderless World*, Harper Collins Publishers Limited, New York

PADCO (1990): *Nepal Urban Development Policy Study*, AID Project no 940-1008.95, Washington D. C.

Siddiqui, Kamal (1992): *Local Government in South Asia: A Comparative Study*, The University Press Limited, Dhaka

Sivaramkrishnan, K. C. and L. Green 1986: *Metropolitan Management: The Asian Experience*, Oxford University Press, New York

UNRISD (1995): *States in Disarray: The Social Effects of Globalisation*, United Nations Regional Institute for Social Development, Geneva

Verma, K. (1998): *Slum Networking Project* (mimeo), Office of the Municipal Commissioner, Ahmedabad

World Bank 1998: *Reducing Poverty In India: Options for More Effective Public Services*, World Bank, Washington DC.

7

NGO and Community Initiatives in Urban Environmental Management

Recent Trends and Success Stories

Nazrul Islam

1. Introduction

Urban environmental management (UEM) envisages management of the environmental aspects of urbanization to achieve environmental policy goals, ultimately leading to sustainable urban development. UEM comprises two concepts: urban environment and management. Both these concepts are complex. The term 'management' is used both in a general sense and in a technical professional sense.

In the context of urban development, the term management (or urban management) has a precedence in (urban) administration while the more recent expression is (urban) governance. Both urban administration and urban management have implications of being functionally authoritarian and techno-bureaucratic top-down processes. Governance, on the other hand articulates a process that is more participatory and inclusive, as it anticipates the participation of both the government (or governmental agencies) and the governed, i.e., civil society. However, in this essay, management is considered to be synonymous with governance, as the latter is the more contemporary and also more acceptable term.

The popularly used contemporary term 'civil society' denotes a broader spectrum, which includes non-governmental organizations (NGOs), community-based organizations (CBOs) citizens groups, professional organizations, etc. The private sector is yet another major actor in urban management. Discussion on the role of civil society in urban management forms a significant part of this paper.

2. Urban Environment and its Components

The urban environment consists of both the physical components and the non-physical components – economic, social, cultural, political and psychological. The paper focuses on the physical environmental components, although the non-physical environmental components are equally important.

The urban physical environment has many subcomponents, which can be further categorized under the natural environmental group and the man-made physical environmental group. The natural/physical urban environmental issues include soil and geology, topography and terrain, soil, hydrology and drainage, air and atmospheric elements, water and aquatic resources, natural hazards like flooding and riverbank erosion, cyclones and tornadoes, earthquakes and landslides, volcanic eruptions, etc.

While the natural/physical environment (except for the occasional natural calamities like earthquakes and floods) does not pose day-to-day problems for urban management; human interventions may lead to degradation of the natural physical environment, raising serious concerns among urban environment managers. These problems include, *inter alia:*

- Incidence of flooding due to the improper management of rivers
- Waterlogging due to the improper construction of embankments or badly planned Urban expansion
- Groundwater depletion and lowering of the water table
- Subsistence of urban land surface
- Water pollution due to discharge of untreated industrial waste water into water bodies
- Shrinking of open spaces
- Noise pollution from transport, industries, commercial activities
- Improper arrangements for solid waste disposal
- Inadequacy of services like water supply, sanitation, sewerage, electricity, road space, etc.

3. The Context

There is widespread concern about the urban environment – from citizens in general to urban authorities, management professionals, development agencies and environmental activists. However, the responsibility for urban management in general, and urban environmental management in particular, rests primarily with the concerned local urban authorities (city corporations or municipalities) and with agencies or organizations at higher levels (state/ provincial or national). These urban local governments and central government authorities, in turn, ensure participation by the private sector and civil society, including NGOs and the community.

Thus, civil society, NGOs and CBOs, normally play a complementary and additional role in urban environmental management. However, if governmental agencies become too ineffective or weak, NGOs and CBOs may initiate corrective measures to improve environmental conditions. In some cases, external agencies and international funding organizations also volunteer to help.

A case in point is Dhaka – the capital city of Bangladesh. The traffic conditions and the solid waste disposal system in this city have become so utterly unsatisfactory that daily newspapers carried reports that the Prime Minister of the country had given directives for monitoring cleanliness and improving civic amenities in the metropolis (*The Independent*, 26 October 2001, Dhaka). The directives covered removal of uncleared garbage and repair of city roads. In another news item on that day, the US Ambassador to Bangladesh was reported to have volunteered help to the Prime Minister in improving the air quality of Dhaka by assisting in the introduction of CNG-run two-stroke autorickshaws in the city (*Prothom Alo*, 26 October 2001, Dhaka). Such are examples of higher level interventions in local urban environment management.

Local urban environmental problems are thus the function of local, regional and global conditions. The various factors influencing the local urban environment include macroeconomic conditions (growth rate, inequality, poverty); national policy (e.g., with regard to city-wards migration, industrial location); legal and regulatory measures; financial policy (e.g., incentives to decentralize industries and offices); measures for awareness creation and motivation of the people and entrepreneurs and leaders. The examples set by leaders (both political and economic) for environmental improvement and better management are also very important.

4. NGO and Community Initiatives

In recent times, NGOs and communities have become involved in many areas of development. These include economic development (employment generation, income enhancement, access to credit, skill development); social development (education and health), political development (human right, rights to vote, etc); cultural development (artistic and literary movements); infrastructure development (building roads, embankments); housing, and environmental management (both of a broad and holistic nature, as well as rural and urban environment).

NGO or community initiatives in urban environmental management relate to various levels/ stages of intervention. These initiatives include awareness creation among and motivation of concerned, citizens, city authorities, national-level functionaries and leaders, running advocacy programmes influencing policy formulation, and implementation of national/ local plans of action.

In urban environmental management, NGOs and CBOs work in partnership with international agencies, national organizations, urban local authorities, elected officials, individual households, ordinary members of the community professionals/ technical persons (with specialized skills and knowledge in field of development intervention), and the media (both print and electronic).

5. Case Studies

Our sources of information on NGOs or community initiatives in urban environmental management are basically (a) publications of UN organizations on human settlement or urbanization, which cover such initiatives; (b) other publications (including journals and occasional papers); (c) unpublished documents in Bangladesh) and (d) personal knowledge (particularly of those in Bangladesh).

Examples of NGO and community initiatives in urban environmental management vary in scope, both in the territorial/spatial sense and in the activity/component sense. Thus, an NGO initiative may cover a whole range of cities, or just a single city or even only a part of a city. Similarly, an NGO may implement only one environmental component or management programme such as solid waste management or air pollution.

While there may be some examples of comprehensive city-scale initiatives or urban environmental improvement by NGOs, numerous examples are available of specific issue-based initiatives by NGOs (ranging from sanitation to park conservation) or small area environmental improvement cases by community-based organizations.

There has been a global trend towards increased involvement of and participation by NGOs and communities in urban management, including urban environmental management. Such participation results both from invitations from the city authorities as well as the initiatives of NGOs and communities. Quite often, such participation is encouraged and made possible by UN programmes or the programmes of foreign (donor) countries or organizations. Many UN programmes make it obligatory for NGOs and CBOs to participate in the design and implementation of the programme. A recent example in Bangladesh is the Sustainable Environmental Management Programme (SEMP) supported by UNDP, which has several urban environmental management activities that must be implemented by NGOs and CBOs. The UNDP–LIFE (Local Initiatives For Environment) Programme is also for small local NGOs or CBOs to implement in urban areas. However, some international organizations suggest that urban environmental management programmes be undertaken by the local urban authority with the participation of NGOs and CBOs. Some NGOs bring their own programmes and work through the local urban authority, either directly or with their assistance.

The success of the NGO and CBO programmes depends on many factors. These include internal motivation of these organizations, their leadership quality, relationship of the NGOs and CBOs with the local urban authorities and the credibility of the organizations with the donors. The attitude of the national government towards NGO activities is also very important. Most NGO/ CBO initiatives begin when a problem reaches crisis proportions; others begin at the time of natural hazards, epidemics or economic disasters.

a. Capacity building support

Most urban local authorities lack adequate capacity to undertake initiatives to improve the environmental quality of their towns.

Box – 1: SINPA – Bangladesh Programme to Implement Habitat Agenda

The programme which is funded by the Netherlands Development Cooperation offers the services of the Institute of Housing Studies (IHS) to help Bangladesh and two other countries Bolivia and Zambia in the implementation of the National Plan of Action for Human Settlements. It aims to provide support by enabling national capacity building institutions to become more responsive to local needs and by maximizing learning both from local and external sources. The programme fully utilizes relevant ongoing projects and programmes to improve the understanding and communication of experiences relevant to the development needs of the city in the linked areas of housing and environment management through participation and partnership. Broadly its core concept is to assist local governments and their partners in capacity building for action in the broad areas of housing, local environmental management and participative processes and partnerships relating to these subjects. SINPA is a programme that supports and values existing initiatives rather than initiating new ones. SINPA's achievements in Tangail Municipality of Bangladesh with the help of the local partner RADOL, include improvements in the solid waste management system, traffic management system, slums and housing facilities for the poor. of the Tangail City, SINPA has also contributed to cleaning the canal which flows in the middle of the city.

Source: Ministry of Housing and Public Works, Government of Bangladesh, 2001: Bangladesh National Report: Progress of Implementation of the Habitat Agenda, 1996-2001

They suffer from inadequacies not only in financial resources but also skilled and motivated personnel. They often require external support to enhance their capacity in financial management, environmental improvement, traffic management or urban planning. Such comprehensive support has been offered to a secondary city (Tangail) in Bangladesh, under the Sustainable Implementation of National Plan of Action (SINPA), for human settlements by the Institute of Housing Studies (IHS), Rotterdam and RADOL, a national NGO in Bangladesh (Box-1).

b. Urban Infrastructure Improvement

Improvements in various infrastructural components (water, sanitation, drainage, sewerage and solid waste management) as a whole is a task taken up by few NGOs. A good example of such an initiative is provided by a federation of 22 NGOs in Kathmandu, Nepal, supported by GTZ. A special feature of the project, which is located in Swavambhu, is voluntary labour from the community. Thanks to the project, public toilets in the area are among the best maintained in Nepal (UNDP-World Bank 1998).

In Dhulikhel, Nepal, a water supply project is being run by the Dhulikhel Water User's Committee since 1981. The project has been successful because the people felt the need for it and the community leader was very committed.

One of the most frequently cited best practice examples in urban low-income area improvement is the Orangi Pilot Project (OPP) in Karachi, Pakistan (See Box 2). OPP has assisted nearly 1 million people in the Orangi settlement to develop their own low-cost sanitation and sewerage programme. While OPP provided the technical support, labour and financing came almost entirely from the community. OPP added health, family planning and income-generation activities to the basic component of the programme, namely sanitation and sewerage.

Box – 2 : Orangi Pilot Project in Karachi

Orangi is an unauthorized settlement, extending over 8,000 hectares with around 1 million inhabitants. Most inhabitants built their own houses but none received official help for the construction. There was no public provision for sanitation; most people used bucket latrines which were emptied every few days, usually on to the unpaved lanes running between houses. More affluent households constructed soakpits, but these filled up after a few years. Some households living near creeks constructed sewage pipes which emptied into the creeks. The cost estimates of the local government agencies to lay sewage pipes in Orangi was too high for local residents, who felt that these pipes should be provided free.

A local organization called the Orangi Pilot Project (OPP) was convinced that if local residents were fully involved, a cheaper, more appropriate, sanitation system could be installed. Research was undertaken by the OPP staff to show inhabitants the serious consequences of poor sanitation on their health and their property. But the residents could neither afford conventional systems nor had they the technical or organizational skills to explore alternative options. OPP organized meetings for those living in 10-15 adjacent houses on each side of a lane, explained the benefits of improved sanitation and offered technical assistance. Where an agreement was reached among the households of a lane, they elected their own leader who formally applied for technical help. Their site was surveyed, plans were drawn up and cost estimates prepared. Local leaders kept their groups informed and collected money to pay for the work. Sewers were then installed, with the maintenance organized by local groups. The scope of the sewer construction programme grew as more local groups approached OPP for help and the local authorities began to provide some financial support. Over the last eight years, households in Orangi have constructed close to 69,000 sanitary pour-flush latrines in their homes, plus 4,459 sewerage lines and 345 secondary drains, using their own funds and under their own management. One indication of the viability of this work is the fact that some lanes have organized and undertaken lane sewerage investments independently of OPP; another is the households' willingness to make the investments needed for maintenance.

Women were very active in local groups, with many being elected as group leaders. It was often women who found the funds to pay for the sewers out of household budgets. But women had difficulty visiting health centres — custom dictates that they should stay at home working through women's groups — as also being active at the level of the lane, where advice is provided on hygiene, nutrition, disease prevention, family planning and kitchen gardens.

Source: UNCHS- An Urbanizing World: Global Report on Human Settlements, Oxford 1996.

c. *Slum improvement*

Urban growth in developing countries has been associated with the proliferation of slums and squatter settlements, notorious for their highly degraded physical environment. Infrastructural investments are needed in roads, drainage, sanitation, sewerage, water supply, electricity, garbage management and fuel supply. The social and economic conditions of the people living in these settlements also need to be improved. There are many good examples of slum improvement. Most of these have been initiated by national governments or local urban authorities, often with support from international donor agencies or UN organizations.

Indonesia's Kampung Improvement Programme (KIP) is one such well-known programme. Jakarta and Surabaya have become very famous cases as both of them were awarded the prestigious Aga Khan Award in Architecture. In Surabaya, the programme was implemented with the active participation of the Surabaya Municipal Authority, the Institute of Technology, Surabaya (acting as the intermediate agency or NGO) and the citizens of Kampung).

Similarly in Kitagata, Japan, an urban environmental improvement programme was implemented in the 1980s for an area (Buraku) inhabited by a backward community, called Burakumins. The intermediary agency in this case was the Wakatake Planning Research Institute. The people of Buraku actively participated in the programme (UNCHS 1997). Examples of university-based groups participating in urban environmental improvement are found in the Sustainable Implementation of National Plans of Action (SINPA), supported by the Netherlands government in Bolivia and Zambia during 1998-2001. In Zambia, the Copperbelt University has worked with the Kitwe City Council in a number of areas for urban improvement, including refuse collection and urban planning capacity building. In Bolivia, an Urban Development Plan was prepared by a group of three universities working with the Santa Cruz City Authority. In Bangladesh, several NGOs work on slum improvement. These include such large ones as Proshika, BRAC and CARE.

d. Solid Waste Collection and Recycling

There are many examples of good practices in the area of NGO/community initiatives in solid waste collection. In Dhaka,

Waste Concern, an NGO, is engaged in household waste collection and composting in a scientific manner. The collection is from both slum and non-slum areas.

The original idea came, however, from a community initiative in Kalabagan. An individual in this middle class neighbourhood undertook the task of collecting household refuse for a small charge and depositing the garbage at a municipal collection point. This financially viable initiative now has many replications in Dhaka and other cities of Bangladesh. In Khulna city, an NGO called Prodipan operates a community-based solid waste collection system in wards covering 5,000 households, assisted by the municipal authority. The method is similar to the one adopted in Dhaka and Tangail.

In India there are many such examples. SWASTI (Solid Waste Initiatives) in Lucknow has been initiated by Muskan Jyoti Samiti. This NGO is currently active in 22 slums of Lucknow. (UNDP-World Bank 1998). The Centre for Development Communication (CDC), a Jaipur based NGO, also has an expanding (covering 400,000 people) programme of solid waste collection for a small fee Rs.20 per household.

Wastewise, in Bangalore, runs an innovative programme that combines waste separation and recycling activities with income generation for street children (Hardoy, Mitlin and Satterthwaite 1992). EXNORA, a local NGO in Chennai, has been involved in a community participatory programme in solid waste management since 1989, in collaboration with the municipal corporation and with financial assistance from RHUDO/USAID. At present, EXNORA collects 600 metric tons (out of the city's 3000 tons) of solid waste. EXNORA involves local communities, called Civic EXNORAs, to start self-help programmes for addressing urban environmental problems.

In Pakistan, the Sindh Katchi Abadi Authority (SKAA) provides support for solid waste management through small communities. The municipal corporation organizes secondary collection and disposal at landfills.

e. *Water supply to slum dwellers*

Normally, Dhaka Water and Sewerage Authority (DWASA), a state- owned agency, does not provide piped water to illegal slum settlements, which are numerous in the city. However, a local NGO, Dustha Sastha Kendra (DSK), has played an intermediary role between DWASA and slum-dwellers at several locations, purchasing water from DWASA for supply to the poor slums. DWASA is gradually formalizing the arrangement, which has functioned satisfactorily.

f. Community-managed sewerage

Sevantha, a well-known Sri Lankan NGO, initiated the management of a local sewer network in an urban slum near Colombo. The self-paying project was jointly undertaken by Sevantha and a formally organized CBO. A women's savings group also participated in the project. High motivation, an effective CBO and good leadership by the NGO made the difficult project a success.

g. Housing Support

Sevanatha has also successfully implemented a participatory housing project in Kandy, Sri Lanka, in collaboration with the local urban authority, the National Housing Development Authority and the local slum community. The basic premise of the project was that the existing shelters should not be destroyed; rather they should be improved upon. The National Housing Policy too played a supportive role. The project was successful because the housing units constructed were very inexpensive and could be afforded by the occupants. In addition, work on the local drainage and water supply system provided employment for the poor residents of the community.

6. Civil Society Movements for UEM

While NGOs and CBOs engage in urban environmental management, awareness creation as well as problem solution and project implementation, most civil society groups perform the role of environmental watchdogs. These groups make citizens aware of urban environmental problems, ill-motivated plans and anti-people projects. They often actively protest against the government or local government policies/measures with adverse consequences for the environments. Neither do the private sector projects with similar negative effects escape their attention.

In India, the most influential civil society institution of this nature is the Centre for Science and Environment (CSE). Among other activities, it has published several volumes of the well-known Citizen's Report on the State of India's Environment. Several other groups have also played an activist role in improving the quality of India's urban environment.

In Bangladesh, the Bangladesh Poribesh Andolon (BAPA) has become a strong environmental movement within a short time (See Box 3). It has vigorously agitated against a number of ill-conceived projects and against encroachment on the Buriganga River and the lakes of Dhaka. It has been campaigning with some success against the severe impact of two-stroke engines on air quality. The Bangladesh Environmental Lawyers Association (BELA), Ain O Shalish Kendra and other legal associations have filed writs against a number of government agencies whose actions have caused damage to the environment of Dhaka and other cities in the country.

The media has also been playing an active role in highlighting urban environmental problems in Bangladesh. *Daily Star*, *Daily Prothom Alo*, *Daily Sangbad* and other major newspapers have organized many important dialogues on such issues. Bangladesh Environmental Journalists Forum, Centre for Urban Studies (CUS), Bangladesh Boy Scouts, Bangladesh Geographical Society, Engineers Institution of Bangladesh, Bangladesh University of Engineering and Technology (BUET) and other civil society organizations have also played advocacy roles for a better urban environment.

With NGOs and civil society groups continuing to be active on environmental issues, the government fully recognizes the importance of involving these groups in national level-committees concerned with environmental policy formulation and project monitoring. Civil society partnerships in environmental management is better understood now than ever before although the distance to be covered is still very great.

Box – 3: Bangladesh Poribesh Andolon (BAPA)/

Poribesh Rokkha Sopoth (POROSH)

BAPA is a civil society organization with clearly declared objective to work on all sectors of the environment but participation of environment-loving people and organizations not taking any donor assistance. POROSH, which had a similar mission, has merged with BAPA. In order to sustain and expand the unity of the pro-environment forces of the country, achieved through the International Conference on Bangladesh Environment (ICBEN-2000); a new platform named Bangladesh Poribesh Andolon (BAPA) was officially launched in June 2000. A vacuum has been filled by this platform. More people are now realizing that environmental issues can be more effectively highlighted if they campaign jointly with BAPA or POROSH. The network and image BAPA has created within such a short period are considered valuable even by the big national NGOs. This could be further strengthened for the movement to reach a goal. Following are two some of the important achievements and recent activities of BAPA.

Buriganga Bachao Andolon: The movement to save River Buriganga has participation by 25 organizations and education institutions and hundreds of enthusiastic individuals. Gradually, the concept of river and wetland conservation is getting the attention of the government, policy planners and city authorities due to people's pressure. Mass media is playing a very key role in keeping the issues alive. In fact as a result of the intense campaigning by BAPA and others, measures have been initiated by the caretaker government since July 2001, and the present government to remove unauthorized constructions along the river.

Air Pollution: As many as 20 organizations and several activists have joined hands to find ways to reduce air pollution. City authorities have also taken cognizance of the problem and BAPA representatives are holding as well as participating in different meetings, seminars, and conferences to reach consensus on the possible action plans.

7. Conclusion

Traditionally urban management, and therefore the management of the urban environment, was the responsibility of city authorities and central government functional agencies operating at the city level. But with growing problems, declining efficiency of city personnel, perennially inadequate finances, and the new concept of urban governance, the roles of the private sector, NGOs, community-based organizations and civil society have become important. Civil society, in particular, can play a major role through its advocacy programmes, as also by keeping pressure on both the central and local governments on the one hand, and the private sector and NGOs — CBOs on the other.

REFERENCES

Government of Bangladesh, 2001, *Bangladesh National Report: Progress of Implementation of the Habitat Agenda, 1996-2001*, Ministry of Housing and Public Works, Dhaka.

Hardoy Jorge E, Diana Mitlin, and David Satterthwaite, 1992, *Environmental Problems in Third Word Cities,* Earthscan Publications Ltd., London.

The Daily Prothom Alo, 26 October 2001. Dhaka

The Independent, 26 October 2001. Dhaka

UNCHS, 1997, *Partnership for Local Action: A Source Book on Participatory Approaches to Shelter and Human Settlements Improvement for Local Government Officials,* UNCHS and CityNet, Dhaka.

UNCHS 1996, *An Urbanizing World: Global Report on Human Settlements,* Oxford.

UNDP-World Bank, 1998, *Water and Sanitation Programme, Notes,* p. 19-20.

8

Urban Environmental Management and the Role of Local Government in India
Some Fiscal Issues

Tapan K Banerjee

Introduction

Urban local bodies (ULBs) in India, as in other developing countries, are perceived as major players in urban environmental management. For, most of the services traditionally being provided and expected to be delivered by ULBs are essentially environmental functions. Thus, for effective urban environmental management, the financial health and management capacity of the ULBs is sine qua non. However, the track record of ULBs in general, presents a dismal picture in this context. This is a matter of great concern.

As the expected domestic and international investments are made in the wake of implementation of the liberalization policy, for achieving the higher rate of industrial growth, there would be faster expansion of economic activities in urban areas because, industrial locations tend to gravitate towards urban centres to take advantage of the infrastructure that is expected to be available there. Thus, the envisioned investments for industrial growth, from both domestic and foreign sources, are conditioned to a large extent by the matching productive advantage that the urban centres can offer by providing adequate infrastructure and services.

Civic services, including those traditionally provided by ULBs, are now perceived as essential "municipal environmental services". Urban environmental protection, as far as these services are concerned, depends on how best these are provided by ULBs, both in quantitative and qualitative terms. It is generally known that there are wide gaps in municipal service provisions, which expose the urban populace to serious environmental hazards. The sustained provision of even basic minimum municipal environmental services for a large urban population, in conformity with specified standards, is albeit a gigantic task. Apart from resource constraints and the inadequate technical and managerial capacity of ULBs, this task is imbued with complexities and intricacies mainly due to the maldistribution of the urban population. The distribution, as it is, is highly skewed towards larger cities, particularly the million plus ones.[1] The population density in large cities has also been alarmingly high.[2] The continually growing urban population and the densification of existing urban centres on the one hand, and the widening gap between demand for and delivery of urban services on the other, have largely contributed to the deterioration of the urban environment and hence the poor quality of urban life. The issues of financing the huge requirement of capital investments and the required inflow of revenue for sustaining them cannot perhaps be addressed by the traditional pattern of financing urban services and the prevailing state-local fiscal relation.

The critical issues that arise in this context are:

- The constraints that prevent ULBs from improving the provision of services in terms of standards and coverage
- Magnitude and dimension of the task ahead for the provision of municipal services of such quality and coverage as would meet environmental criteria
- Available innovative policy options for augmenting the fiscal and financial capacity of ULBs so as to match their functional responsibilities with resource availability
- State-local fiscal and functional relations that would aid the ULBs in performing their assigned functions
- Extent to which the 74th Constitution Amendment Act(CAA), 1992 can address the above issues

These issues will be briefly addressed in Part I of this paper, to sketch the current state of municipal services in general across the country. This is necessary for understanding the background of the local fiscal issues that need to be resolved for bringing about structural reforms in municipal finance in India. The broader objective is to enable ULBs to provide such services as would meet the necessary conditions for environmental protection. This will be followed by a quick assessment of the perspective of the demand-based task ahead of ULBs. Then a brief review of the present status of municipal finance across the country will be made, reflecting on the functional and fiscal devolution to ULBs by states. Part II will focus on the new innovative options that have been adopted or could be explored by some states and ULBs for enhancing municipal financial and management capacity. Finally, a broad reform agenda that addresses the critical fiscal and related issues, will be suggested.

PART I

The Background

Local bodies, both urban and rural, is a state subject according to Indian constitution. A municipality (including municipal corporations) performs such functions as are assigned to it by the respective state legislature. While some of these functions are obligatory, others are discretionary in nature. The legislature also gives municipalities the power to impose specified taxes and to raise revenue from specified non-tax sources. The purpose of these financial provisions in the municipal acts is to match the functional assignments to the municipalities with financial resources made available to them. Thus, in the past (pre 74th CAA), it was the states' prerogative to create a municipality, supersede a municipality for an indefinite period, delegate financial power, devolve functional responsibilities and so on. Though 'local bodies' is still a state subject, the 74th CAA has ushered in a new era in the history of the institution of urban local government. Besides giving constitutional recognition to municipalities as institutions of local self-government, the CAA incorporates some landmark provisions with the underlying objective of democratization of ULBs, decentralization of urban management responsibilities down to the grass-roots level and institutional strengthening of ULBs. It is beyond the scope of this paper to review the CAA per se. However, specific provisions will be referred to as related issues are discussed in the paper.

There are18 functions listed in the Twelfth Schedule that the 74th CAA desires urban local governments to perform. One of these functions is "urban forestry, protection of the environment and promotion of ecological aspect". Though discretionary unless the state legislature makes these functions obligatory by law, this constitutional provision reflects the consciousness in respect of environmental degradation that the city-dwellers are threatened with.

The 74th CAA does not codify the measures or actions that ULBs should take or for that matter the services that they should provide to prevent environmental degradation. However, a number of functions listed in the Twelfth Schedule may be construed as "environmental services". These include:

- Regulation of land use and construction of buildings
- Water supply for domestic, industrial and commercial purposes
- Public health, sanitation, conservancy and waste management
- Urban forestry, protection of environment and promotion of ecological aspect
- Slum improvement and upgradation
- Urban amenities and facilities such as parks, gardens and playgrounds
- Burials and burial grounds; cremation, cremation grounds and electric crematorium
- Public amenities including street lighting, parking lots, bus stops, and public conveniences
- Regulation of slaughterhouses and tanneries

ULBs have traditionally been providing these environmental services assigned to them by the states, though not being able to fully respond to the minimum civic need. There are instances of some major municipal services being provided by parastatal (or quasi-government) authorities. But that does not change the basic nature and character of these services. They essentially remain municipal services; or for that matter, urban services, irrespective of who provides them.[3] With growing concern about the urban environment, most of the municipal functions or services are now perceived, as mentioned before, essential "municipal environmental services".

Gaps in Municipal Services

There is a plethora of literature that deals with the gaps in municipal services, which expose the urban populace to environmental hazards. Information readily available in this regard is a bit dated. Nevertheless it provides broad indicators (national averages) of the state of some of the municipal services, particularly those that gives rise to environmental concern. These indicators as listed below give some idea about the magnitude and dimension of the problems.[4]

Water supply

- 81.59 per cent urban household had access to drinking water (1991).
- 53.30 per cent had access within the premises; 40.13 per cent within half a km (1988-89).
- Supply of water per capita varied between 165 litres per capita per day(lpcd) in Class I cities to 54 lpcd in Class IV towns (1995).
- Average water supply in urban slums was 27 lpcd while the minimum standard fixed by the National Drinking Water Mission was 40 lpcd (1995).
- In lakh plus cities, 61 per cent of the water drawn was from surface water sources, which were generally contaminated (1982).

Sewerage and sanitation

- About 48 per cent of the urban population had access to latrines of one kind or other; of them, 23.45 per cent used service latrines (1992).
- The population covered by the sewerage system varied between 75 per cent in Class I cities to 35 per cent in Class IV towns.

- About half the waste water generated is collected and one-fourth is treated; untreated raw sewer flows directly to surface drains, creating alarming environmental hazards.
- Untreated industrial waste contaminates both surface water and groundwater.

Solid waste management

- In large cities, on an average, 60 per cent of the solid waste generated is collected everyday. For smaller towns, the average is 50 per cent.
- Most solid waste collected ends up in open dumps or drainage systems, threatening both surface water and groundwater quality.

Drainage

- Storm water drainage service covers only 66 per cent of the urban population.
- In most of cases, rainwater, solid waste, sewage and sludge from domestic and commercial sources are all flushed away by a common drainage system. In addition to posing serious health hazards, this causes flooding of roads due to choked drains.

Roads

- Due to the bad state of repair and maintenance, most of the municipal roads are generally reduced to dusty tracks, leading to air pollution, besides the inconvenience faced by users.

Since the beginning of the Seventh Plan, considerable initiative both at the national and state levels has been taken to expand the civic amenities in quantity, quality and coverage. However, these efforts could not keep pace with the growing urban population and the increasingly higher demand for civic amenities. Neither could interventions for urban development adequately address the problems of city slums – about which the Seventh and the subsequent plans, and several commissions and committees expressed their concern. The gaps in the provision of municipal environmental services widened with the continual growth of urban population.

In terms of the degree of urbanization (percentage of urban population to total population), India cannot claim to be a highly urbanized country. Yet it happens to be one of the largest urban systems in the world because of the sheer size of its population. A mere 26.8 per cent of the total population works out to an urban population of 285.35 million in 2001. The annual growth rate slowed downed from 3.1 per cent in 1981-91 to 2.7 per cent in 1991-2000. However, in absolute terms, the urban population increased sizably – from 62.4 million in 1951 to 285.35 million in 2001, i.e., an addition of 222.95 million in five decades.[5]

The chronic financial constraints of the municipalities prevent them from adequately responding to the growing need for municipal environmental services. In fact, the ULBs' requirement of financial resources for municipal environmental management has increased manifold for reasons beyond the demographic issues. As a result of increasing industrial and economic activity in urban centres, it is expected that the personal income of the urban working population will grow at a faster rate than it has previously. This, in turn, will lead to accelerated demand for municipal environmental services.

Thus, availability of financial resources is a critical issue. The additional financial resource requirement may be categorized into two groups:

(a) *Revenue fund* for meeting:

i. Current deficit in the revenue budget

ii. Additional recurrent commitments arising out of investments made to wipe out the backlog in environmental services

iii. Commitments emerging out of new infrastructure created in view of the demand from new industries and associated expansion of commercial services

(b) *Capital fund* for

i. Creating assets to wipe out the current backlog in environmental services

ii. Investments in civic infrastructure in order to create the necessary environment for facilitating economic, or for that matter industrial growth in urban areas

Issues in Financing Municipal Environmental Services

The state-local or for that matter intergovernmental fiscal relations need to be viewed against the background of the pattern of function sharing between the higher and lower echelons of government. For, "clarity in the division of functional responsibilities between levels of government is an essential condition for any reform in the structure of urban service delivery."[6] Essentially, there should be a guiding principle for this functional divide. In terms of the theory of fiscal federalism, urban services may be categorized into three groups based on the level of government responsible for their provision:[7]

a) *Typically municipal functions*

These are functions whose benefits do not generally extend beyond the territorial jurisdiction of a municipal body and thus need to be entrusted to municipal bodies.

b) *Transmunicipal functions*

The geographical coverage of the benefits of these services cuts across municipal territorial jurisdiction. These services are the natural responsibility of the higher level of government.

c) *Agency functions*

Although they are the responsibility of the higher level of government, these functions can be entrusted to the local government on behalf of the former on an agency basis.

"In India, many [typically] local services such as water supply, sewerage and.... are provided [in most of the states] by autonomous agencies or departmental undertakings and not by the urban local bodies."[8] The theory of fiscal federalism or for that matter any guiding principle is hardly applied for determining the pattern of sharing functional responsibilities between the state and urban local governments in this country.

Rather it is guided by such policy decisions as are taken by the state governments from time to time. This state of affairs is reflected in the interstate disparity in the functional domain. For example, in a number of large cities and metropolises, in particular, typically local functions like water supply and sewerage have been entrusted to para-statal organizations by taking them out of the functional domain of the municipal body. Again, there are wide interstate and inter-ULB disparities in the standard of services provided and their geographical coverage. This makes for complexity, which is not conducive for devising a uniform policy in the context of the state-local fiscal relationship.

The financial powers delegated to ULBs, coupled with the state government's financial assistance to them, should match their functional responsibilities. For, it is the state legislature that delegates, by law, the functions that the ULBs are obliged to perform. In this context, it may be worthwhile to take a look at Dillon's Rules. These state, "It is a general and undisputed proposition of law that a municipal corporation possesses and can exercise the following powers, and no others: first, those granted in express words; second, those necessarily or fairly implied in or incidental to the powers expressly granted; and third, those essential to the accomplishment of the declared objectives and purposes of the corporation – not simply convenient but indispensable."[9] Depending on what the state legislatures provide by law, ULBs in some states of India have a long list of obligatory functions, while in others, most of the municipal functions are discretionary. In any case, irrespective of their legal obligations, ULBs are expected to perform the "typically local functions". From another point of view, it is basically the state government that is responsible for the provision of all urban services; the ULBs share as much of this responsibility as the government wants them to. An imperative of this state-local functional relationship is that the state government so empowers the ULBs as to enable them to strike a balance between revenue and expenditure in the municipal budget.

The resource base of ULBs in India is generally composed of:

1. Recurrent receipts

i. Own source revenue

a. Tax revenue:	Property tax; other taxes
b. Non-tax revenue:	Fees, tolls, service charges, etc; user charges; miscellaneous receipts
ii. Shared revenue:	Entertainment tax; stamp duty on transfer of immovable properties; motor vehicle tax
iii. Grants from state governments:	Dearness allowance grant; compensatory grant in lieu of abolition of local sources, like octroi and entry tax; other specific purpose grants

2. Non-recurrent receipts

i. Capital grants from the state government for specific purposes

ii. Loans: from the government; from financial institutions; from the debt market

In India, the local tax base in all the states is narrow. In most states, property tax is the mainstay of municipal own source revenue. The five states where this is not the case are Maharashtra, Gujarat, Rajasthan, Orissa and Punjab. In these states, octroi is the single largest revenue earner. The relative contribution of other taxes to the municipal own source revenue is too small to be reckoned with. Except for property tax, there exists reckonable interstate variation across the country in respect of the composition of the municipal revenue. Table 1 gives some idea about this disparity. While most states have abolished octroi, which though lucrative is considered to be an "obnoxious" revenue base, it remains the most important source of own source revenue in the five states mentioned above. ULBs in non-octroi states are in general more dependent on revenue transfers from state governments.

Table 1: Major Sources of Tax Revenue of Some Municipal Corporations in India (1991-92)

States	Municipal Corporations	Sources of Tax Revenue
Maharashtra	Greater Bombay (Mumbai)	Octroi, property tax, wheel tax
West Bengal	Calcutta (Kolkata)	Property tax, profession tax (now taken over by the state government), advertisement tax
Karnataka	Bangalore	Property tax, advertisement tax
Orissa	Bhubaneswar	Octroi, property tax
Gujarat	Surat	Octroi, property tax, vehicle tax, theatre tax
Tamil Nadu	Madras (Chennai)	Property tax, profession tax, advertisement tax
Andhra Pradesh	Hyderabad	Property tax
Uttar Pradesh	Kanpur	Property tax, advertisement tax, terminal tax

Sources: Budgets of the respective municipal corporations, 1991-92.

Non-tax sources of revenue of ULBs, – composed of user charges and fees including miscellaneous charges – constitute a small portion of the municipal own source revenue (see Tables 2 and 3). While fees are a regulatory device, user charges have high revenue potential and are instruments for rationing civic services and hence preventing wastage at the recipients' end – a necessary condition for ensuring accountability of the service providers. In most of ULBs, user charges are yet to be introduced. Even those ULBs that collect user charges (e.g., for water supply) are generally reluctant to revise the water tariff upward. To cite an example, in Karnataka, the government revised water rates by an executive order in 1996. It did so on the strength of the fact that ULBs are, by law, empowered to set and revise water rates in that state. The government had to take this step because the ULBs showed little inclination for revising the water rates for more than a decade.

As can be discerned from Tables 2 and 3, ULBs are increasingly becoming dependent on transfers (grants and shared revenue) from the state governments to bridge their budgetary gaps. Between the years 1993-94 and 1997-98, the state average of municipal own source revenue as a percentage of total municipal revenue declined in all states barring a few. These were Orissa, Punjab, Gujarat and Rajasthan – all octroi-collecting states. There are, however, some exceptions to this pattern. Despite being an octroi-collecting state, this percentage declined in Maharashtra; and in the non-octroi collecting state of Tamil Nadu, this percentage went up. In Maharashtra, this reversal of the general trend can be explained by the decline in the growth of property tax. In Tamil Nadu, municipal fiscal performance steadily improved as a result of the mobilization of revenue from various tax and non-tax sources.

Growth of municipal own source revenue relative to total revenue between 1993-94 and 1997-98 was much slower in all the states covered by the review, with the exception of the non-octroi state of Tamil Nadu and the octroi-collecting states of Gujarat, Orissa, Punjab and Rajasthan. This suggests that the growth of revenue transfers from state governments has been much faster than that of municipal own source revenue, particularly in the non-octroi states.

The tendency to centralize the process of revenue collection from local bases has led to the increasing dependence of ULBs on transferred revenue. The revenue loss of ULBs due to the abolition of a local revenue base (e.g., octroi) or due to the removal of tax bases from the municipal fiscal domain (e.g., profession tax), is compensated by a "grant in lieu" of the abolished tax. The in lieu grants, as past experience has shown, never match the buoyancy of the lost revenue bases and are found to be precariously low. Nevertheless, dependence of the ULBs on transferred revenue has been continuously increasing. This trend is more pronounced in the case of states where octroi,

Table 2: Aggregate Total Revenue and Own Source Revenue of Municipalities in Selected States (Excluding Municipal Corporations) in 1993-94 and 1997-98

States	1993-94			1997-98			% Increase 1997-98 over 1993-94	
	Total Revenue	Own Source Revenue	Per cent	Total Revenue	Own Source Revenue	Per cent	Total Revenue	Own Source Revenue
Andhra Pradesh	193.94	82.20	42.38	321.55	150.19	46.71	65.80	82.71
Gujarat	181.45	109.14	60.15	341.11	184.44	54.01	87.99	68.99
Karnataka	110.23	78.64	71.34	137.21	68.52	49.94	24.47	(-) 12.27
Kerala	72.22	43.44	60.14	175.19	72.02	41.11	142.58	65.79
Madhya Pradesh	62.92	28.32	45.01	113.03	32.46	28.72	79.64	14.61
Orissa	41.03	24.56	59.86	53.53	36.87	68.88	30.47	50.12
Punjab	136.54	102.60	75.14	230.67	196.97	85.39	68.94	91.98
Rajasthan	52.04	44.80	86.09	123.39	89.26	72.34	37.11	99.24
Tamil Nadu	138.80	92.89	66.92	265.78	152.81	57.49	91.48	64.51
All India	1,833.32	1,128.59	61.56	3,272.35	1,925.90	58.85	78.49	70.65

Source: Report of the Eleventh Finance Commission, 2000-05, Annexure viii. 3D. Government of India.

Table 3: Aggregate Total Revenue and Own Source Revenue of Municipal Corporations in some States (1993-94 and 1997-98)

(Rs. in crores)

States	1993-94			1997-98			% Increase 1997-98 over 1993-94	
	Total Revenue	Own Source Revenue	Per cent	Total Revenue	Own Source Revenue	Per cent	Total Revenue	Own Source Revenue
Andhra Pradesh	158.50	78.86	49.75	319.44	190.02	59.49	101.54	140.96
Gujarat	445.22	324.66	72.92	850.99	625.86	73.54	91.14	92.77
Karnataka	135.00	59.00	43.70	257.32	83.29	32.37	90.61	41.17
Kerala	39.22	30.00	76.49	85.09	41.94	48.82	119.02	39.80
Madhya Pradesh	204.22	110.19	53.96	292.63	104.82	35.82	43.29	-4.87
Orissa	20.72	17.00	82.05	36.80	31.12	84.57	77.61	83.06
Punjab	126.30	103.70	82.11	221.96	196.53	88.54	75.74	89.52
Rajasthan	77.92	65.05	83.48	160.22	141.37	88.23	105.62	117.33
Tamil Nadu	247.33	139.12	56.25	630.03	424.88	67.44	154.73	205.41
All India	3899.28	3056.49	78.39	8090.49	5349.03	66.12	107.49	75.01

Source: Report of the Eleventh Finance Commission, 2000-05, Annexure viii. 3D. Government of India.

the most important revenue earner, has been abolished. Two glaring examples are Karnataka and Madhya Pradesh. An NIPFP(National Institute of Public Finance and Policy) study[10] indicates that in 1992, the percentages of revenue transfers to total municipal revenue in octroi states like Gujarat, Maharashtra and Punjab were 10, 5 and 11, respectively, while in non-octroi states like Madhya Pradesh and West Bengal, they were 41 and 56, respectively.

Shared revenue in India relates to motor vehicle tax, entertainment tax, stamp duty and the like. In most states, the shares of the ULBs were fixed long ago and are rarely adjusted with the price index. Thus, in real terms, their share has been declining over the years.

ULBs are free to utilize the portion of the transferred revenue that comes in the form of shared revenue and in-lieu grant. They do not enjoy autonomy in spending so far as other forms of transferred revenue or grants are concerned. These are tied or specific-purpose grants. In West Bengal, a substantial portion of the grant is earmarked as compensation for the impact of inflation on the municipal wage bill – commonly known as "dearness allowance grant". In other states too, ULBs get this grant in various proportions. Other examples of specific-purpose grants are education grant and preventive health care grant and flood relief grant. In some states, a large number of specific purpose grants of relatively small amounts are given to ULBs. Maharashtra is one such an example.

Notwithstanding the growing dependence of ULBs on transferred revenue, there is no transparent policy or well-pronounced principle for the devolution of grants, which in most of the states, constitute a substantial portion of the transferred revenue. In the state budgets or otherwise, there is no discernible divisible pool of revenue earmarked for devolution to the ULBs. In this respect, ad hocism prevails.

State Finance Commission

There were high expectations that the recommendations of the State Finance Commissions (SFCs), made on the basis of their terms of reference, would resolve most of the aforesaid issues. It was also hoped that in the context of a constrained resource position, the ULBs would be so rehabilitated as to be able to meet their present obligation towards providing civic amenities as also the obligations that would arise with the envisaged growth of the secondary and tertiary sectors of the economy.

Under Article 243- I of the 74th CAA, each state is obliged to constitute every five years a state finance commission, "which shall review the financial position of the municipalities [and panchayats] and make recommendations to the Governor for strengthening the financial position of the municipalities." The guidelines or the terms of reference for the SFC have also been provided under this Article of the CAA.[11] The 74th CAA has also amended Clause (3) of Article 280 of the Constitution, which requires the Central Finance Commission to take "the measures needed to augment the Consolidated Fund of a State to supplement the resources of the Municipalities in the State on the basis of the recommendations made by the Finance Commission of the State". This is a landmark constitutional provision in the history of local finance in India. It is a step forward towards integrating the recommendations of the SFCs with that of the Central Finance Commission and the terms of reference of the SFCs indicate to forging a healthy state-local fiscal relationship.

The first SFCs in all the states submitted their recommendations three to five years ago and the second ones have been constituted in most of the states. It was expected that the recommendations of the SFCs would bring about substantial changes in the composition and structure of local revenue in the states, and in the state-local financial and functional devolution. However, no reckonable changes were discernible. Regarding the sharing of resources, the SFCs of some states recommended the earmarking of a portion of the state revenue, something like a divisible pool, for devolution to the ULBs. Different basis or formulae were also recommended for working out the entitlement of each ULB to the divisible pool. The proportion of state revenue to be earmarked for devolution varied widely from state to state. Some examples are given below.[12]

State	SFCs' Recommendation for Revenue Sharing
Andhra Pradesh	39.24 per cent of the state tax and non-tax revenue
Karnataka	5.4 percent of the state's gross non-loan receipts for meeting the plan and non-plan requirements
Kerala	1 per cent of the state revenue (excluding some sources) as non-plan and non-statutory transfer to urban and rural local bodies on the basis of population
Madhya Pradesh	8.67 per cent of the state's tax and non-tax revenue
Punjab	20 per cent of the net proceeds of five taxes; namely, stamp duty, motor vehicle tax, electricity duty, entertainment tax and cinema shows
Rajasthan	21.8 per cent of the net proceeds of state taxes to be devolved to rural and urban local bodies in the ratio of 3.4:1
Tamil Nadu	8 per cent of the state's net tax receipts in 1997-98; with the percentage gradually increasing in successive years to 9, 10, 11, reaching 12 per cent in 2001-02
Uttar Pradesh	7 per cent of the state's net proceeds from all taxes to be transferred to urban local bodies
West Bengal	16 per cent of the net proceeds from all taxes (except some minor ones) to be devolved to rural and urban local bodies on the basis of population and some other parameters

Source: State Finance Commission Reports of respective states

The recommendation for creating a divisible pool, however, did not make a significant addition, if at all, to the municipal revenue. As shown in Tables 5 and 6, in aggregate terms, the majority of states had deficits in the budgets of municipalities and municipal corporations in the year 1997-98. This hardly reflects any impact of the recommendations of the SFCs on the municipal finance even three to five years after the Commissions submitted their reports. Many of the state governments were not too enthusiastic about implementing the recommendations of the SFCs; delays and partial implementation of the recommendations were not uncommon. What the SFCs (in some states at least) could at best achieve was initiating a process of rationalization of the transfer mechanism in respect of grants and other variety of transfers. The SFCs of all the states did not, however, take this approach. Again, barring the recommendation regarding the divisible pool, those made by all the SFCs were ad hoc in nature and failed to address the lacunae existing in the state-local fiscal and functional relationships.

Table 5: Revenue, Expenditure and Revenue Gaps of Municipalities in Some States 1993-94 and 1997-98

(Rs. in crore)

States	1993-94			1997-98		
	Total Revenue	Total Expenditure	Revenue Gap	Total Revenue	Total Expenditure	Revenue Gap
Andhra	193.94	171.33	22.61	321.55	291.88	29.67
Gujarat	181.45	191.05	-10.05	341.11	369.61	-28.05
Karnataka	110.23	93.75	16.48	137.21	131.24	5.97
Kerala	72.22	85.12	-12.09	175.19	177.98	-2.79
Madhya	62.92	70.27	-7.35	113.03	140.08	-27.77
Orissa	20.72	53.08	-33.08	36.79	62.42	-25.63
Punjab	126.03	136.54	-10.24	221.96	230.67	-8.71
Rajasthan	77.92	52.04	25.88	160.22	123.39	36.83
Tamil Nadu	247.33	200.48	46.85	630.03	393.63	236.04

Source: Report of the Eleventh Finance Commission, 2000-05, Annexure viii 3D, Government of India.

Table 6: Revenue, Expenditure and Revenue Gaps of Municipal Corporations in Some States – 1993-94 and 1997-98

(Rs. in crore)

States	1993-94			1997-98		
	Total Revenue	Total Expenditure	Revenue Surplus/Gap	Total Revenue	Total Expenditure	Revenue Surplus/Gap
Andhra	158.50	214.32	-55.82	319.44	412.84	-93.40
Gujarat	445.22	327.30	-13.85	850.99	369.61	-481.38
Karnataka	135.00	148.85	117.92	257.32	280.41	-23.09
Kerala	39.22	51.43	-12.21	85.09	94.64	-8.74
Madhya	204.22	243.65	-39.43	292.63	398.25	-105.62
Orissa	20.72	21.99	-1.27	36.08	38.71	-1.91
Punjab	126.03	131.47	-5.17	221.96	202.79	1917
Rajasthan	77.92	80.58	-2.66	160.22	157.61	2.61
Tamil Nadu	247.33	315.47	-68.14	630.03	563.06	66.97

Source: Report of the Eleventh Finance Commission, 2000-05, Annexure viii 3D, Government of India.

PART II

Financing Municipal Capital Improvement Works

Given the issues discussed in the forgoing sections, the ULBs' capacity to finance civic amenities and, more importantly, the needed infrastructure to cope with the envisaged growth in the urban economy, remains a big question. So far, there has been no uniform pattern of financing civic amenities by ULBs across the country. The only commonality discernible among them is the cumulative backlog in the provision of civic services, both vertical and horizontal, due to the chronic financial crunch. Certainly there are some exceptions, but they are few and far between.

Traditionally, municipal capital improvement works have been financed primarily by the budgetary allocations of the central and state governments, which municipalities received both as grants and loans. Besides, smaller municipalities obtained loans from Life Insurance Corporation (LIC) of India for water supply schemes. Recently, Housing and Urban Development Corporation (HUDCO) has emerged as a major financial institution, providing loan assistance to municipalities for capital investment in civic infrastructure. Loans from these financial institutions are required to be backed by government guarantees. Even the budgetary allocations and institutional finance from HUDCO and LIC put together have proved far too inadequate to meet the need for investable capital in respect of the huge backlog in civic infrastructure development.

New Initiatives

From the perspective of infrastructure (including urban and civic infrastructure) as a priority sector of development, various options are now being considered or are available for financing municipal capital works. Some of these are:

- A central/state government loan fund that includes assistance from bilateral and/or international donors

- Loans for urban infrastructure from state-controlled institutions, like pension fund and insurance fund
- State-sponsored financial intermediaries (like in Tamil Nadu, Kerala, Karnataka and Gujarat) that have been created for leveraging grant funds with loan funds, thereby, reducing the overall cost of borrowed funds to the municipalities
- Direct access of municipalities to the debt market for financing infrastructure projects by means of "general obligation bonds" or "revenue bonds"
- Pool Finance Development Fund of Government of India, proposed to be created for providing credit enhancement to facilitate pooled finance borrowing by small and medium ULBs
- Private sector participation (PSP) in urban infrastructure development and municipal service provision, which has several variants, such as:
  - ✓ Build-Operate-Transfer (BOT)
  - ✓ Build-Operate-Own (BOO)
  - ✓ Build-Own-Operate-Transfer (BOOT)
  - ✓ Build-Own-Lease-Transfer (BOLT)
  - ✓ Franchising
  - ✓ Contracting out the management of a service
  - ✓ Concession
  - ✓ Public-private partnership (joint venture)

There could be other ways in which capital funds can be mobilized for developing urban and civic infrastructure and services.

Private sector participation (PSP) in the provision of civic services in the form of BOT, BOO, joint venture or other methods is welcome. For, it would reduce the financial liability and management load of ULBs to the extent of PSP in infrastructure provision. Inevitably, PSP will entail a more realistic pricing of the services that are provided, in order to raise revenue. There is an apprehension that in the wake of PSP, the urban poor will be priced out and the profit motive may overshadow the social objectives of municipal service provision. To guard against this, an effective *regulatory mechanism* needs to be put in place. Nevertheless, the question that still needs to be answered is: Will there be enough takers of civic services provided by the private sector? There has been PSP in some civic service provision, but such examples are few and far between.

It would be an ideal situation if municipalities could develop civic infrastructure all by themselves via loans or debt finance. Borrowing per se is not an issue in this context. The critical issue is the present and potential capacity of ULBs to sustain the investments made out of borrowings. In other words, do ULBs have the required debt-carrying capacity? Will they be able to meet the operation and maintenance (O & M) commitments related to the assets emerging out of the new investments, in addition to their existing liabilities? Even when there is no borrowing and the entire development work is grant-funded, the additional O & M commitments are still required to be met out of municipal revenue fund. Some schemes, like water supply, can be self-financing; that is, the marginal revenue generated by such a scheme will at least equalize the marginal cost of the scheme. This subsumes pricing of the service provided by the scheme. Here again, the question of social objectives arises. Bringing the urban poor within the fold of the pricing mechanism may not be socially just and politically acceptable. Thus, the scheme can either be partly financed out of the general revenue and partly by user charges or by setting an enhanced tariff for those who can afford to pay, to cross-subsidize the scheme in favour of those who cannot afford to pay for the service. Even this method of financing cannot be applied to most municipal services, as the benefits that an individual avails of from their provision cannot be measured.

The aforesaid limitation of user charges notwithstanding, it is an effective instrument for the full or partial cost recovery, as may be appropriate in a particular situation, of such services as may be priced. Most of the development expenditure cannot but be financed out of general municipal funds. Only in the case of a few services (e.g., water supply), the benefits to an individual can be quantified. There are a few others that can be commercialized (e.g., recycling of solid waste). Full cost recovery of even such services by means of user charges or commercialization does not seem to be a feasible proposition. As the things stand today, there is no complementarity between user charges, taxes and other municipal own source revenue, and transferred revenue (grants and shared revenue). The solution lies in a package of reform measures embracing all of these sources of revenue.

Agenda for Reforms: Financial Capacity Building of ULBs

Bahl and Linn[13] have suggested a framework for the identification of resources that are appropriate for financing different types of municipal expenditure:

- Where local services are measurable and accrue to readily identified individuals within a jurisdiction, user charges are the appropriate financing instrument
- Local public services such as administration or street lighting, which are services to the general public — in the sense that identification of beneficiaries and measurement of benefits and costs to individuals are difficult — are most appropriately financed by taxes on local residents
- The cost of services for which significant spillovers to neighbouring jurisdictions occur, e.g., health, education and social welfare, should be borne substantially by state or national intergovernmental transfers
- Borrowing is an appropriate source to finance capital outlays on infrastructure services, public utilities and roads

Within this simple framework, a tentative agenda of reforms is suggested below:

1. Typically local functions should be entrusted to the ULBs by law, restricting the power of the state government to remove such functions from the municipal domain.
2. All functions listed in the Twelfth Schedule of the 74th CAA should be made, by law, obligatory for the ULB.

3. Functions that are entrusted to the ULB but whose benefits extend beyond the municipal territorial jurisdiction should be entitled to grants from the state government, at least to the extent of the spillover to the neighbouring areas.
4. Local resource bases should match the local functions entrusted to the ULB, so that the full cost of operation and maintenance of services can be met out of local resources.
5. To the extent possible, shared taxes and specific-purpose grants may be replaced by block grants out of a pool fund to be created by the state government on a rational basis of devolution, as has been recommended by some SFCs.
6. Despite being the mainstay of the municipal own source revenue in India (except in octroi states), the potentiality of property tax has not been exploited so far. Its base needs to be maintained so as to give it buoyancy without of course, sacrificing the equity aspect.
7. To make property tax more productive, all its four aspects – identification of properties, record keeping, assessment and collection — need to be improved. A broader base is necessary for its long-run growth.[14]
8. The system of self-assessment of property tax, like in the case of income tax, may be introduced.
9. ULBs may be empowered by law to set the rates of property tax without any conditions imposed by the state government.
10. Economic rates, needed to break even the cost of providing a municipal or environmental service, should form the basis for determining user charges. The question of affordability of the poor may be addressed by a transparent policy of subsidy for them.
11. Borrowing, preferably long-term and imposition of user charges, wherever possible, may be resorted to for financing capital expenditure.
12. A clear system of accountability in respect of the decisions taken at the municipal level needs to be developed.

13. In order to improve municipal administration, a management information system (MIS) is urgently needed. For this purpose, a computer-supported data retrieval system needs to be introduced.
14. For bringing about transparency in financial affairs, the present single-entry system of book keeping should be replaced by the accrual-based double-entry system with computer application. Additionally, the entire budgetary practice requires to be recast.
15. The citizenry should be taken into confidence before a major decision is taken by an ULG. Constitutional ward committees or similar grass-roots level organizations need to be activated for this purpose.
16. A time frame may be set, perhaps by means of legislative measures, for the implementation of the recommendations of the SFCs by the state governments.

This agenda, obviously, is not comprehensive. Each item has its ramifications. Should this agenda be agreed in principle, details of actions can be worked out.

NOTES AND REFERENCES

1 Only 23 urban agglomerations and million plus cities together, i.e. about 0.70% of 3301 ULBs and urban agglomerations across the country, accounted for 31% of the total urban population in 1981. The same number of urban agglomerations and million plus cities i.e. around 0.65 of 3697 ULBs and urban agglomerations accounted for 32% of the total urban population.

2 For example, the density of population per square kilometer in Delhi, Kolkata and Mumbai in 1991 was 13,486, 12,282 and 10,697 respectively. See Urban Statistics – Hand Book 2000, National Institute of Urban Affairs, New Delhi.

3 Encroachment of the municipal functional domain by the state governments, which goes much against the policy of decentralization is a debatable issue, not addressed in this paper.

4 Singh B N, 1999,"Financing Urban Infrastructure Development in Local Government" paper presented at International Seminar on Financing and

Pricing of Urban Infrastructure, 8-10 February 1999, New Delhi, HSMI, HUDCO, New Delhi, pp. 162-164.

5 Census of India, 2001, Series 20, West Bengal, Provisional Population Totals, Distribution of Urban-Rural Population. Government of India, New Delhi

6 Mathur O P, 2001, Approach to State-Municipal Fiscal Relations: Options and Perspectives, National Institute of Public Finance and Policy, New Delhi, p 7.

7 Mohanti P K, 1995, Reforming Municipal Finances: Some Suggestions in the Context of India's Decentralization Initiative, p. 8 (mimeo.)

8 Ibid.

9 Dillon John F 1991, Commentaries on the Law of Municipal Corporation, Little Brown Boston.

10 National Institute of Public Finance and Policy 1995, State-Municipal Fiscal Relations: Options and Perspectives for the State Finance Commissions, vol. II, Supporting Tables, NIPFP, New Delhi.

11 (a) The principles which should govern the recommendations of the Commission:

i. The distribution between the state and the municipalities (and panchayats) of the net proceeds of the taxes, duties, tolls and fees leviable by the state, which may be divided between them under this Part and the allocation between municipalities at all levels of their respective shares of such proceeds;

ii. The determination of the taxes, duties, tolls and fees which may be assigned to or appropriated by the municipalities;

iii. The grants-in-aid to the municipalities from the Consolidated Fund of the state;

(b) The measure needed to improve the financial position of the municipalities;

(c) Any other matter referred to the Finance Commission by the Governor in the interest of sound finance of the municipalities.

12 Mathur, State-Municipal Fiscal Relations and SFC reports of states.

13 Bahl RW & J F Linn, 1992 Urban Public Finance in Developing Countries, Oxford.London

14 Ibid.

9

Urban Environmental Issues in India
Approaches to Environmental Planning and Management

T M Vinod Kumar

Introduction

This paper attempts to present a very broad canvas of urban environment management issues that have emerged in India in recent years. This provides an insight into multi-level policy making and appropriate urban environmental planning and management interventions. Undoubtedly, the statistics being collected in India are inadequate for town-level environment planning and management. What is required is a town web site that annually updates environmental statistics for effective urban environment management at least up to the ward level for all Class I towns to start with. The paper discusses several urban environment management approaches practised in India, such as rapid urban environmental assessment, environment mapping, environmental management plans and environmental zonation. These can lead to useful solutions if equal importance is given to the urban environment and urban economic development. Neither should be privileged over the other.

The 74th Constitutional Amendment Act (CAA) of India has strengthened the hands of urban local governments and communities (wards committees) in various areas, including environment management. However, not many states in India have utilized its provisions in the area of urban environment management effectively for the benefit for the local population. Individuals, local communities, community-based organizations (CBOs), non-government organizations (NGO) and urban local bodies (ULB) jointly have the power to manage the urban environment. This can be done within the framework of the annual planning and budgeting system of municipalities supported by the state and central finance commissions under the 12th Schedule of the 74th CAA and various other legislation, including municipal acts. These bodies, communities and organizations have the potential to transform the urban environment so that it fosters healthy living. The availability of finance has always been a constraint on municipalities' attempt to find solutions to mammoth urban environmental problems, which have roots in non-promotion of local initiatives, a centralized administrative set-up and large-scale neglect over five decades or more. Therefore, the implementation of environment management solution calls for the mobilization of additional organizational and financial resources by local governments with the active participation of NGOs, CBOs and local communities.

In addition, it is important to build the capacity of these institutions to monitor and mange the urban environment. The job involved is technical in nature, but very much within the capability of local CBOs, NGOs, and ward committees. State-of-the-art approaches practised in India will be discussed in Section 3 of this paper. Further effort is required for the development of appropriate training modules and programmes on ward level environment monitoring, planning and management, which is beyond the scope of this paper.

India has a range of environmental NGOs. There are specialists NGOs, whose focus of interest could be hospital waste, urban forestry, watershed management or water conservation; as also generalist NGOs with a broad spectrum of interests. The NGOs can be divided roughly into three groups: *(a)* technologically disenchanted "greens" concerned about the "absence of power to transform" the nation in the direction of self-sufficiency, equitable resource distribution and traditional culture; *(b)*"eco-developers" anxious to shift development toward ecologically sustainable paths; and *(c)* environmental managers, who seek cost-effective environmental approaches. They are critical to environmental management throughout India, but their reach at the local level is limited. They increasingly engage in issues of technological change and urban and industrial affairs. Their voice is heard via judicial activism, including highly effective legal action through public interest litigation.

Urban environmental issues, which will be discussed in the following section of this paper, first become evident to the local community. Therefore, the local community should have the capacity to identify such issues and find remedies for environmental maladies, using their constitutional powers and access to municipal budgets. They should also be empowered to mobilize additional finances and organizations, if required. Effective urban environment management (UEM) in India calls for creating partnerships between local governments, NGOs/CBOs and local communities. Decentralization is the key to addressing urban environment issues. The commitment of the state government to decentralized planning and management at the town, village and district levels is the foundation for this effort. While some environmental problems can be solved through local participation, others need to be tackled globally or nationally. The latter can be identified locally and debated for national and international consensus on management.

2. Indian Urban Environmental Issues

2.1 Urban Environmental Conditions

Health conditions of urban inhabitants and their economic productivity in India have been significantly impaired by environmental degradation. A World Bank study concluded that India's annual environmental damages conservatively amounted to \$9.7 billion or 4.5 per cent of the gross domestic product at 1992 values, compared to 2.6 per cent for China. Approximately, \$7 billion of the estimates are due to water and air pollution (World Bank 1996a). The following facts illustrate the problem:

The three big metros in India – Delhi, Kolkata (formerly Calcutta) and Mumbai – are among the 10 most polluted cities in the world. Perhaps, the situation is no better in Mexico City, Rio de Janeiro, Bangkok and Tokyo. Six of India's largest cities experience very high levels of air pollution. Bombay, Kolkata, Delhi, Ahmedabad, Kanpur and Nagpur have annual average total suspended particulates that are at least three times the World Health Organization (WHO) standards. Respiratory ailments such as asthma are common. Reductions in the levels of suspended particulates to conform to the WHO standards would save an estimated 37,000 lives annually. A stock of ancient automobiles and growing numbers of two-stroke, two- and three-wheelers exacerbate the problem. In India, the burning of biomass, wood and coal for domestic cooking/heating in urban slums, is another contributory factor in air pollution. The daily pollutant load in Indian cities has shown a rising trend. This is likely to increase the incidence of several illnesses, with the attendant costs of treatment. Mounting air pollution in Indian cities leads to a greater prevalence of dense smog in winter. In addition, there is a worsening of the problem of global warming. To this, India is contributing 8 per cent, China 10 per cent, Western Europe 17 per cent, and Eastern Europe with Russia 16 per cent, Africa 6 per cent, Australia 3 per cent, U.S. 33 per cent and South America 8 per cent (US-AEP 1996). The rapid rates of population growth in poor countries is also adding to the rising CO^2 levels. The photochemical smog, supersonic aircraft flights and use of refrigerant gases are depleting the stratospheric ozone levels. Among the long-term effects of such depletion is an increase in the rates of cancer of skin and liver, cataracts and dermatitis. Thus, the adverse impact of urban, industrial air pollution extends far beyond national borders. Its effect on health over the next 20 years will be quite dramatic. Clearly, reduction in the levels of air pollution should be given high priority (Ibid.).

Economic use of scarce surface and groundwater to meet the basic needs is perhaps the most important challenge facing India today. For example in Delhi, wherever drinking water is pumped from underground, neither the quantity nor the pressure is sufficient to lift it to overhead tanks because of falling groundwater levels. Poor quality and maintenance of water supply networks results in leakage of scarce water, contributing to low water availability in summer. Large government investments in surface storage, drinking water supply and irrigation, along with the increasing private mining of groundwater, have continued since Independence. Yet rivers that are dry, except during the monsoon, because of upstream dams now only receive sewage or industrial waste. Widespread salinity, waterlogging, siltation, pesticide and fertilizer pollution in rural areas, and industrial pollution pose serious health risks. Obtaining potable or industrial water is a daunting challenge. As of 31 March 1997, 90 per cent of the urban population had access to water supply facilities; three years later, the figure fell to 88 per cent, pointing to the deteriorating conditions (HUDCO and Habitat 2001). These percentages, however, hide the quality and adequacy of urban water supply services in India.

Water supply networks are old, poorly maintained and generally of indifferent quality, resulting in high transmission and distribution losses, ranging from 25 per cent to 50 per cent. Low pressure and intermittent supplies allow back siphoning, which results in contamination of water carried by the distribution networks. Untreated sewage is the primary source of water pollution. Statistic reveals that as of 31 March 1997, 49 per cent of urban population had sewerage and sanitation facilities. The figure rose to 55 per cent over the next three years. Out of 300 Class I towns with populations equal to or more than 100,000, about 70 per cent have partial sewerage systems and sewage treatment facilities. Sewerage network is provided in less than 20 per cent of India's largest cities; even here the facilities are partial. The level of sewage treatment is very low. Consequently, untreated or partially treated municipal waste water finds its way into precious water sources such as rivers, lakes and ground water, causing water pollution (Ibid.). For example, about 75 per cent of the waste water that flows into the Ganges comes from municipal sources.

Industrial effluent is largely generated by the 3 million small- and medium-sized units that are scattered throughout the country, particularly during the production of paper, sugar, leather and chemicals. Only about half the medium- to large-scale industries have partial or complete effluent treatment facilities. The fourfold industrial growth during 1963–91 resulted in a sixfold growth in toxic releases. This increased further over the last decade. Industrial chemical and iron and steel producers contribute nearly 70 per cent of the toxics released but only 20 per cent of the industrial output. Industrial disposal of polluted effluent occurs via open drains into streams and reservoirs or through underground injection. Most industrial estates lack waste water treatment systems (US-AEP 1996).

Solid waste collection and disposal systems in many cities are severely strained. As per estimates, about 100,000 metric tons of municipal waste is generated daily in India. Per capita waste generation ranges from 200 to 600 grams per day. The collection efficiency is 70 to 90 per cent in metro cities and 50 per cent in smaller cities. The urban local bodies spend about Rs 500 to Rs.1,500 per ton for solid waste management. Of this, 60 to 70 per cent is spent on street sweeping and collection, 20 to 30 per cent on transportation and even less than 5 per cent on final disposal (HUDCO and Habitat 2001). Most cities do not keep pace with the build-up of waste. Storage facilities for non-hazardous industrial waste (such as fly ash) and safe disposal of hazardous waste do not exist. Although some companies are able to recycle or incinerate these wastes effectively, hazardous waste output is increasing from chemical, textile and other industries (Sumathi 1993, Frank 1984, CEE 1993). The investment requirement for appropriate solid waste management initiatives (including biomedical waste management for urban centers with different population groups) is estimated as follows (HUDCO and Habitat 2001):

Table 1
Investment Requirement for Solid Waste Management

Population (100,000)	Rs (million)
1	40
5	80
10	130
20	24

Sound pollution is rampant in urban areas. Indian cities are noisy places. During religious ceremonies and marriages, blaring mikes add to the already high noise levels. Delhi motorists don't feel happy without frequently blowing their horns, whether or not it is required. At traffic intersections, they use their horns to the maximum.The decibel stress at most intersections is way above the permissible levels, which can impair hearing. Industries also aggravate the problem of sound pollution.

2.2 Urban Health and Environment

The link between environment and health is being rediscovered. During early 19th century, cholera was recognized as a waterborne disease. However, this was followed by a medical view that stressed the curative rather than preventive aspects of disease. Social and environmental conditions were merely seen as contributory factors to the spread of disease-bearing pathogens. In the late 1960s, the "book of infectious diseases" was declared closed in the North. Less than 20 years later, old and new diseases emerged – rendering even the best health care systems inadequate in both the North and the South. The process of development itself appears to contribute to the spread of disease. Unplanned urbanization, and the resulting deterioration in basic services, has increased respiratory and gastrointestinal infections in urban areas. A shift in thinking from curative to preventive health is now unavoidable. It should emphasize primary health care in urban centers as was done in rural areas.

2.3 Urban Environment and Infrastructure

It is now recognized that having adequate urban infrastructure in place is vital for accelerated economic development as also for a healthy living environment. As elsewhere in the world, cities in India are the engines of economic growth. Along with the central government, state and municipal governments are playing an increasingly important role in developing infrastructure such as water supply and waste water, solid wastes, and hazardous waste disposal systems. However, these initiatives are far from sufficient to meet the existing gaps and growing demands. The *India Infrastructure Report*, popularly known as Rakesh Mohan Committee Report, estimated that infrastructure investment requirement would increase from about Rs 4,000 – 4,500 billion (US $ 115 to 130 billion) during 1995-2000 to about Rs 7,500 billion (US $ 215 billion) during 2001-06(Ministry of Finance GoI 1996). This investment requirement is staggering. With India experiencing unprecedented recession and deflation, the only scope for kick-starting the national economy and increasing the GNP growth rate is to increase the investment in urban infrastructure development, which has considerable backward and forward linkages.

The major plan schemes under urban infrastructure include: (*a*) Integrated Development of Small and Medium Towns; (*b*) Mega City Infrastructure Development; (*c*) Special Component Plan for the National Capital Region Planning Board; (*d*) Accelerated Urban Water Supply Programme, (*e*) Low-cost Sanitation; and (*f*) Urban Transport (Ministry of Finance, GoI 2002).

The India Infrastructure Report had estimated an annual requirement of funds of Rs 2,80,000 million for providing water supply and sanitation facilities to the urban population. Compared to this, the total allocation for schemes pertaining to urban water supply and sanitation during the Ninth Five Year Plan (1997-2002) was Rs 1,86,240 million – a huge short fall in investment (Ministry of Finance, GoI, 1996).

The central government is planning to augment the financial resources available for waste water projects by allowing privatization. However, the current waste water market is limited, with no private operation and management opportunities. Basically, an underdeveloped market exists for wastewater services throughout India. Municipalities are still at the micro-level, problem-solving, fire-fighting stage – attempting to improve the management of these basic urban services. However, there is some promising departure from this general trend, such as Tiruppur in Tamil Nadu where young industrialists came forward to tackle the issues of water supply and sewerage with modern management and private capital. Tiruppur is a thriving industrial town specializing in underwear garments and exports the products internationally. Alandur Sewerage Project in Tamil Nadu is another unique experiment of public participation in project financing. In 1996, Alandur municipality, 14 km near Chennai, approved a proposal to provide an underground sewage/drainage system for a targeted population of 300,000 to be completed in five years at a capital cost of Rs 337.54 million. Loans (59 per cent,), grants (12 per cent), deposit collection from participants (23 per cent), and interest and deposits (6 per cent) were used to finance the project. About Rs 77.5 million has been spent on various activities, and the project is progressing successfully (Mathur 2002).

The Ministry of Environment and Forests has provided funding to several cities to conduct surveys on urban solid waste disposal. It was found that landfills are costly due to restricted land availability. Most non-hazardous, non-organic solid waste (plastic, tires, metal, glass, etc.) is sold as scrap, which is then recycled. In terms of guidelines, rules and regulations, a very important development took place during the Ninth Plan. In March 1999, a report on solid waste management of Class I cities was submitted to the Supreme Court of India. A mid-term appraisal of the Ninth Five Year Plan states that management of municipal solid waste is one of the most neglected areas of urban development in India. Lack of financial resources, inadequate manpower, fragmentation of administrative responsibility, and non-involvement and lack of awareness of the community are some of the major reasons for the inadequate attention. Like waste water, solid waste management is currently an underdeveloped market with little privatization. In solid waste management, privatization is limited to small contracts on a short-term basis in several municipalities. There are exceptions in the form of a dozen compost plants and a common treatment facility for biomedical waste management in Hyderabad. In 2000, Chennai permitted private sector participation in solid waste management (only collection and transportation) in about 33 per cent of the city with the help of an international company. There is scope for private-public participation in solid waste management at a much larger scale.

2.4 Clean Production and Environmental Management in Urban Centers

Many urban centers in India are transforming themselves into hubs of industrial production. They are the "back offices" of large cities in the country. The quality of life in these centers will suffer unless a policy of clean urban industrial production is followed. An important input that can influence policy formulation in this direction is public awareness and participation. There is a strong evidence of heightened awareness about and pressure for a clean environment from the non-governmental community via public interest litigation in the Supreme Court. Concern for sharing information with the people at large has struck a strong chord in India. The wide availability of information on pollution discharge and the impact and risk assessment is crucial for public debate on environmental issues. Making such information available to an even wider section of the interested public in a systematic and sustained way can greatly strengthen the demands for moves toward ever cleaner technology and production processes. It could also help address the concerns about the environmental impact of large corporations as well as the actions (or inactions) of government agencies. Urban local bodies with the participation of NGOs, engineering colleges and local communities should take up the environmental monitoring of municipalities and industries.

Despite analytical support from the Central Pollution Control Board and sister organizations at the state level, there is pervasive cynicism and pessimism about the effective implementation of the existing environmental regulatory (i.e., command and control) and management systems in India. Local bodies with the support of the central and state governments should offer fiscal incentives to promote environmental quality at the firm/industry level in urban settings.

3. Appropriate Urban Environmental Intervention

3.1 Rapid Urban Environmental Assessment

As a methodology, rapid urban environmental assessment (RUEA) draws its inspiration from rapid rural appraisal and participatory rural appraisal. The former, developed during the 1970s, is a fairly quick and clean technique for development planning. It seeks to avoid unsuccessful agricultural projects that follow the top-down and blueprint approaches to rural development. Participatory rural appraisal, developed during the 1980s, involves data collection, analysis, problem identification, ranking of opportunities, preparation of village-level resource management plans and follow-up. However, urban assessment is much less anthropological and community-focused than its rural counterparts, primarily because cities involve much larger populations and spatial areas. In summary, the methodology of RUEA consists of a three-step process:

- Completion of a questionnaire on urban environmental data
- Preparation of an urban environmental profile, using data from the questionnaire and research assistance from local investigators
- Discussion of the results through a series of consultations, culminating in a priority-focused public workshop

The three-step process was adopted to enable local experts and citizens to rapidly assess the state of the urban environment. This process is based on measurement, observation, validation and action. This is elaborated as follows:

- To *measure* a consistent set of data, an urban environmental data questionnaire is to be used.
- To *observe* the nature, trends and factors that influence environmental quality in cities, a common framework for preparing an urban environmental profile should be developed.
- To partially *validate* and use the results from the questionnaire and the profile as a basis for follow-up action, consultations with key actors in the cities are to be held. Most important, these consultations are a means of arriving at a consensus and developing the political momentum to act on priority problems.

Subsequent steps in this approach include:

- the formulation of an integrated *urban environmental management strategy,* which embodies issue-specific strategies, long-term environmental goals and phased targets for meeting the goals;
- agreement on *issue-oriented action plans* for achieving the targets, including the identification of least-cost project options, policy reforms and institutional actions; and

- a *consolidation* phase in which agreed programmes and projects are initiated, policy reforms and institutional arrangements are solidified, the overall process is made routine, and monitoring and evaluation procedures are put in place.

The phases need not follow a strict sequence. For a given city, one might start at the beginning, middle or end of the process delineated above, depending on the existing level of consensus on environmental priorities, as well as the political and socio-economic context.

Local governments of urban or urbanizing localities are best suited to use RUEA as they provide a comprehensive diagnostic approach to identifying problems as well as solutions for a wide range of environmental issues. Local planning bodies may particularly steer its implementation. RUEA can also be utilized to find solutions based on the 10 principles of New Environmentalism described later in this paper.

3.2 Three Components of Environmental Planning

Environmental planning seeks to improve and protect environmental quality for urban residents – both by controlling the generation of pollution and by segregating activities that are environmentally incompatible (Miller and Groot 1997). Effective environmental planning requires regular interaction between three components – hardware, software and "heart-ware" (see Figure. 1).

Figure. 1

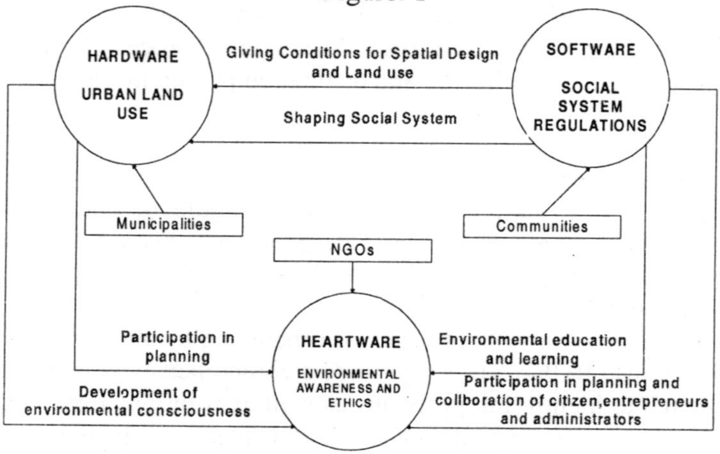

Source: Adapted from Harashina (1996).

- *Hardware* consists of the physical fabric of a city – infrastructure, buildings, railway, roads, ports, etc.
- *Software* consists of the rules, and regulations, laws and legislation, ordinances, habits and ethics, and traditionally established codes of conduct.
- *Heart-ware* refers to an individual's emotional make-up, behaviour and conduct, which determine his or her needs, wants and wishes. It also determines the process by which the individual's interacts with the surrounding environment.

It is within the above framework that the contribution of Central Pollution Control Board (CPCB) of India which has adopted the following multipronged approach to urban environment management will be discussed.

3.3. Environmental Mapping

Environment base maps: The preparation of an environmental management plan requires a series of maps to understand the area and identify the trends. The characteristics of the urban area, viz., land use, drainage, topography etc., provide the base structure for deriving the components/sectors that are subsequently analyzed. Reference maps are the most important information bases for understanding the spatial dimensions. The base map is used as the reference map, which forms the base of each of the theme maps. The base map includes the urban area (study area) boundary; major roads/railways; names of important settlements; major rivers/water bodies; and major landmarks.

Environment hot spots: A map showing all the environmental hot spots affected by air pollution, water pollution, solid waste disposal, land use incompatibility, etc., is prepared, indicating the sources, so that recommendations can be made accordingly. It is very important to identify the sources of environmental pollution and degradation. Otherwise, decisions may be taken wrongly by targeting controls at a wrong source.

3.4. Environment Management Plan

The environmental management plan includes:

(a) Mitigation measures and identification of abatement infrastructure
(b) Land use controls

3.4.1 Mitigation measures and abatement infrastructure

Mitigation measures should be taken for contaminated sites, polluted areas and polluting industries, while pollution abatement infrastructure should be planned for industrial clusters and industrial estates. The most common remedial measures are containment technology, soil washing/fixing, pump and treatment technology, thermal treatment, bioremediation, incineration etc. The mitigation measures for an industry should target controls at source and waste minimization. Recycling, reuse and renovation should be adopted. There should be segregation of streams based on effluent characteristics and application of the most optimal treatment – one that has the flexibility to take care of variations in effluents (type, quality, frequency of discharge) and still meets the required standards.

Typical infrastructure requirements include a drainage system for effluents/sewage; storm water drainage system; solid waste/hazardous waste disposal facility; incinerator for hazardous waste; common effluent treatment plant (CETP); and a green belt. The combined facilities, such as CETP for new industrial estates, should have a modular design since all the industries in an estate do not come up simultaneously. In addition, the facility should be capable of handling a wide variety of effluents/wastes.

3.4.2 Land use controls

The areas of incompatible land use should be identified and, accordingly, the source or the receiver should be shifted. This may include relocation of industries, commercial areas, wholesale markets, truck or bus or railway terminals, roads, railway lines, etc. The incompatibility may also be avoided by providing a buffer between the source and the receiver.

In major agglomerations, where the population exceeds 10 million, the dynamics of various activities and their inter-relationships are quite complicated. This is especially so where a good spatial plan and planned development is lacking. These urban centres require:

- A plan that earmarks the areas for industrial development and specifies the type of industries and the pollution load that can be permitted
- A transportation network that allows free movement of traffic and easy dispersion of emissions
- Green belts, buffers and open spaces

Land use controls around pollution sources are extremely important. If sensitive use is allowed in close proximity to pollution sources, the adverse impact might still be felt notwithstanding the adoption of adequate pollution control measures for meeting the required standards. Land use should be planned according to the type of pollution source. Barring Kanpur and Haldia, not much progress has been made on the preparation of environmental management plans by the CPCB.

3.4.3 Scale of urban environmental issues for management

A key to effective policy formulation and allocation of resources to project components is the understanding of the *scale* of urban environmental problems. While certain environmental issues are specific to an area and can be tackled locally, others need to be addressed regionally or nationally. This understanding will help in ascertaining the appropriate level at which data and information is to be collected; resources and stakeholders are identified; policy is formulated and appropriate action taken. At each spatial scale, the characteristic problems need to be addressed and the related infrastructure and services must be provided. Consideration of the spatial scale of environmental impact reveals several important issues for developing countries (see Figure. 2).

Figure. 2

Water pollution	Loss of habitat, biodiversity and endangered species	**Region/ Nation**	Soil erosion and increased salinity	Toxic run-off and acid rain
Amenity loss Traffic congestion **City** Loss of heritage and historical buildings Reduced property and building values Accidents and disasters Polluted land **Community** Inappropriate and inadequate technology use Trash dumping **Household** Garbage generation, air/water/noise pollution, spread of diseases Lack of understanding of environmental problems Flooding Noise pollution Natural disasters Inadequate tax/financial revenues Flooding and surface drainage				

Lack of inappropriate, laws and legislation Toxic and hazardous wastes/dumps High living densities Loss of agricultural land and desertification Air pollution Water pollution Inadequate supply and transmission loss of electricity Misguided urban governments and management practices		
Natural and man-made hazards and disasters	Land clearance and loss of forest cover	Effects of climate change and global warming.

Among the major issues is developing a local information base for urban environmental planning and management. Rapid urban environment assessment is an appropriate approach for building the capacity of local governments, NGOs and ward committees.

4. Urban Industrial Environmental Management

Technology transfer: Too little information is available at the municipal level about technological alternatives for urban and industrial environment management despite several promising opportunities that can be tapped in the laboratories of the CSIR (Council for Scientific and Industrial Research), IITs (Indian Institute of Technology) and engineering colleges. Although India has a large capability for science and technology research and development and a substantial capital goods industry; capacity utilization in these areas is far from satisfactory. Technology transfer in terms of sales to end users is important but insufficient. Signal opportunities exist for promoting collaborative research and joint ventures between educational institutions and equipment suppliers. There is also potential for integrating India into the global technology system through links with the North as well as the South to derive mutual benefits.

Environmental principles for industrial growth: India would benefit immensely from the ability to gauge the intensity of industrial pollution with respect to environmental capacity and government enforcement. Preparation of zoning atlases at the district level was initiated by the CPCB to address this issue; the effort was extended to cover industrial estates. Firms can apply these standards and formulate zonal plans to make informed technology choices and monitor their own performance and compliance.

Voluntary standards: Perhaps the most important impetus for environmental quality within India's industrial regime will be pressure from the ISO 14000 international environmental standard (particularly as it affects export industries). Although some activity to inform industry about environmental management systems is already occurring within industry associations, stronger national support is necessary to make ISO 14000 an effective incentive. L & T Construction major in India has been getting ISO 14000 certifications for many of their projects. Counterparts for collaboration can be found in the government's standards agencies, e.g., the Indian Standards Institution (ISI) and the Quality Council of India; and among business groups, the Association of Chambers (Assocham), Confederation of Indian Industry (CII), Federation of Indian Chambers of Commerce of Industry (FICCI), and other industry-specific organizations.

Greening the supply chain: The movement to "green" supply chains reflects a related pressure from international standards favoring clean production. This development reflects the incorporation of environmental concerns within quality standards, the movement to organize manufacturing networks, and the globalization of business.

Environmental consulting: The Bank of Baroda is actively seeking to improve its environmental consulting capabilities. The principal counterparts for collaboration may be found among specific banks (e.g., Bank of Baroda), industry associations (e.g., banking and insurance industry associations), educational and training organizations (e.g., those institutions specifically geared to the financial sector), and business groups (e.g., the Assocham, CII and FICCI). Vigorous efforts by lending institutions to improve their understanding of environmental risks and liabilities could be an important impetus to increased public disclosure of environmental information by the government.

Expanding markets: The market for water, waste water, and solid and hazardous waste infrastructure in India is expanding at an astronomical rate. The central and state governments remain the major buyers and driving force behind this market. Government credit supports it. Rising public and NGO awareness of domestic pollution and its adverse impact on health will impel increasing government as well as private investment in environmental infrastructure projects.

Management needs of municipalities: The responsibility for environmental infrastructure development will fall on municipalities as per the constitutional amendment. The allocation of funds by state finance commissions for this purpose is meager and inadequate. Throughout India, there is a growing need to build the municipalities'capacity to plan, finance, operate and maintain new infrastructure. Innovative financing approaches through urban land management and other methods, such as municipal bonds and property tax reforms, need to be put into operation. Municipalities also need continuous up-to-date information and guidance on the types of technologies available appropriate to local conditions and the levels of service required.

5. Finding Solutions: The Ten Principles of New Environmentalism

There has been a growing change in the way environmental policies, programmes and projects are formulated and implemented. Among the key features of this new thinking are the link between economic activities and environmental problems; the need for a multiplicity of actors and actions at various levels to solve these problems; and the need to tackle cross-border, global environmental problems at the international level.

Specifically, new environmentalism – characterized by greater precision in factoring environmental costs and benefits into policy making – puts local people at the center of environmental strategies, diagnoses and addresses the behavioural causes of environmental damage, and recognizes the political dimensions of environmental reforms. The ten principles that characterize new environmentalism are described in Table .2 below:

Table 2

Principle 1: Set priorities carefully	The seriousness of environmental problems and the scarcity of financial resources have required tough prioritization and phasing of remedial actions. Broad, shallow and expensive approaches have to be avoided, with serious priority-setting exercises grounded in proven studies and analyses.
Principle 2: Make every rupee count	Unnecessarily expensive environmental policies, based on high-cost approaches traditionally used in industrial countries, have to be avoided. There is a need for a new emphasis on cost-effectiveness, allowing for much more to be achieved with limited resources. It requires a multidisciplinary approach - one that calls for environmental specialists and economists to work together to identify the lowest-cost methods of addressing key environmental problems.
Principle 3: Harness "win-win" opportunities	Some gains in the environment will involve costs and trade-offs. Others can be achieved as by-products of policies designed to improve efficiency and reduce poverty. Reducing subsidies on the use of natural resources and clarifying and reallocating property rights are examples of win-win policies, which are expected to have a beneficial impact on the environment, particularly given the scarcity of resources devoted to solving environmental problems.
Principle 5: Economize on administrative and regulatory capacity	In many countries, administrative and enforcement capacity is usually weak. They recognize that highly "enforcement-intensive" approaches cannot be adopted and therefore emphasize self-enforcing policies and other instruments (such as taxes, bans, fees, etc.) with less intervention. This approach has also increased the role of NGOs and community groups.
Principle 6: Work with the private sector, not against it	Recognizing their limited regulatory capacity and the need for accelerated private investment, many governments are switching from a control-dominated attitude towards the private sector to one that involves

	dialogue and negotiated, monitored programmes. Self-enforcement and independent certifications schemes (for example the ISO14000) are now playing a much larger role.
Principle 7: Involve citizens thoroughly	When a country's environmental problems are addressed, the chances of success are greatly enhanced if local citizens are involved. Such involvement is needed for four reasons: • Local citizens are often better able than governments to identify the priorities for action. • Members of local communities often know about cost-effective solutions that are not available to governments. • The motivation and commitment of communities are often what sees an environmental project through to its completion. • It helps build constituencies for change as a counterweight to vested interests.
Principle 8: Invest in partnerships that work	Often, the most effective way of dealing with environmental issues is when the stakeholders work in partnership. NGO involvement in priority-setting exercises and tripartite relationships (between the government, the private sector and community organizations) are becoming increasingly common. The value of such partnerships stems from not only the different perspectives and skills that are brought to the table, but also the necessity of carrying out concerted actions to address environmental issues.
Principle 9: Remember that management is more important than technology	The old-fashioned, technology-driven approach to the environment is giving way to a recognition of the crucial role of good management. Improved management practices are always a complement to, and sometimes a substitute for, investment in equipment.
Principle 10: Incorporate the environment from the start	When it comes to protecting the environment, prevention is much cheaper, and more effective, than cure. Most countries now seek to assess and mitigate potential damage from new infrastructural investment. Efforts to move such concerns "upstream" in the project cycle, to factor environmental concerns into sectoral strategies, are underway.

Source: Steer (1996).

6. Conclusion

The paper provides a broad overview of urban environment management issues confronting India in recent years. The insights gained can serve as useful inputs for multi-level policy making and appropriate urban environmental planning and management. No attempt was made to marshal the environmental statistics available for Indian cities. National Sample Surveys collect data on urban environmental infrastructure, but the sample size does not allow it to be useful at the district or town level. The Centre for Monitoring the Indian Economy also compiles infrastructure statistics (CMIE 2000). However, the statistics being collected in India are inadequate for town-level environment planning and management. What is required is a town web site that annually updates environmental statistics for effective urban environment management at least up to the ward level for all Class I towns to start with. The ministries of Environment and Forest, Urban Development and Poverty Alleviation and the central and state pollution control boards cannot on their own take on the task of regular data collection, monitoring and updation for the town web site. In this endeavour, they require the active and continuous participation of local engineering colleges, NGOs and CBOs, ULBs and, above all, the participation of local communities.

The control and command regime of environmental legislation and rules and regulations was briefly discussed. Although their importance cannot be underestimated, these instruments, which have existed in some form or the other since the 18th century, have not created a healthy urban environment in India. ULBs should have the jurisdiction to operate environment courts in urban centres. Perhaps they can be in closer proximity to the affected population than the overburdened courts in distant district, state or national capitais.

According to the *India Infrastructure Report,* mammoth investment is required for urban infrastructure. It is not possible to meet the demand for urban environment investment via the older approaches. Commercialization is strongly advocated by this report. Public-private partnerships must be encouraged and capital markets reforms be implemented expeditiously. There are a few notable examples of the former, such as the Tiruppur and Alandur Sewage projects in Tamil Nadu. Many build-operate-and-transfer projects, such as urban roads and bridges, point towards the feasibility of commercialization. This seems to be is the only way out.

The paper discussed several urban environment management approaches practiced in India, such as rapid urban environmental assessment, environment mapping, and environmental management plans and environmental zonation. These can lead to solutions if equal importance is given to urban environment and urban economic development. Urban environment management activities require capacity building. Training modules must be prepared and all stakeholders trained to achieve the best results.

REFERENCES

Centre for Environment Education (CEE) 1993, *Study of City-Wide Best Practices in Solid Waste Manangement in Collection, Transportation and Disposal*, CEE, Bangalore.

Centre for Monitoring Indian Economy (CMIE) 2000, *Infrastructure Statistics Report*, Mumbai.

Frank Flintoff 1984, *Management of Solid Waste in Developing Countries,* World Health Organization, Regional Office of South East Asia, Delhi.

Harashina Sachihiko 1996, *Environmental Planning on Urban Level*, Discussion Paper 96-6, Dept. of Social Engineering, Tokyo Institute of Technology, Tokyo.

Housing and Urban Development Corporation (HUDCO) and United Nations Centre for Human Settlements (Habitat) 2001, *The State of Indian Cities 2001*, HUDCO and Habitat, Nairobi-New Delhi.

Mathur Mukesh 2002, "Alandur Sewerage Project: A Unique Experiment of Public Participation in Project Financing", in *India Infrastructure Report 2002: Governance Issues for Commercialization*, Oxford University Press, New Delhi.

Miller Donald et al 1997, *Urban Environmental Planning,* Avebury Publishers.

Ministry of Finance, Government of India 1996, *The India Infrastructure Report: Policy Imperative for Growth and Welfare*, Expert Group on the Commercialization of Infrastructure Projects; published by Rakesh Mohan, on behalf of MoF GoI, New Delhi.

Ministry of Finance, Economics Division Government of India 2002, *Economic Survey 2001-2002,* Mof GoI, New Delhi.

Sumathi.S 1993, *Solid Waste Management: Best Practices Adopted in Madras City*, Centre for Environment Education (CEE), Bangalore.

Steer Andrew 1996, "Ten Principles of the New Environmentalism", *Finance and Development*, December, pp. 4-7.

United States-Asia Environmental Partnership (US-AEP) 1996, *Environmental Infrastructure Strategy: India*, Draft, *US-AEP/USCS* Washington DC.

World Bank 1996, *India's Environment: Taking Stock of Plans, Programmes, and Priorities: An Assessment of the Environment Action Programme, India*, World Bank, Washington DC, South Asia Regional Office, India.

10

Participation as a Prerequisite for Urban Development

The Significance of Grassroots Democracy for Urban Settlements of the Future

Siegbert Heid

1. Introduction

The 21st century could well turn out to be the century of the urban settlement. For the first time in the history of mankind, the majority of the people will be living in town and cities. The report on the proceeding of the conference "Urban 21" held last year in Berlin put forth the prognosis that the number of urban-dwellers will swell to 5 billion by the year 2025. This would amount to a 47 per cent increase, constituting over 61 per cent of the global population. As per this prognosis, there will be some 358 cities with populations of over a million by the year 2025. Of these, at least 153 will be located in Asia. Eighteen of the 27 mega cities with at least 10 million inhabitants each are already in Asia – among them Mumbai with over 27 million, Kolkata with over 17 million, approximating the population of Delhi, followed by Lahore and Hyderabad with just under 11 million each.

But it is not urban growth per se that will pose the greatest problem, rather the lack of competence, capacities and capital to regulate this growth. To remedy this, something more needs to be done. Agenda 21 of the Rio Conference of 1992 on Environment and Development has set the course which is characterized by a careful and judicious management of our resources. Chapter 28 of the Agenda is devoted to the options before municipalities. When the Rio Conference and the Habitat II Conference of1996 in Istanbul confirmed the social, ecological and town-planning experiences of the inhabitants of particularly underdeveloped cities and parts thereof, optimists believed that the "sustainability of towns and cities" would now advance on the triumphant path to asserting itself. Even though, admittedly, this process has not taken off as if fired by a missile, it is indeed advancing, if rather slowly. Today, sustainability has become an integral part of many urban development programmes. This shows that the concept has a central role to play, particularly at the local level. It is at this level that the citizen displays the greatest willingness to engage actively.

2. Local Democracy: The Global Dimension

But in today's world, local democracy and civil society no longer restrict themselves to the limits of a town or municipality. Action taken at the local level is influenced by global considerations. In other words, there is a growing need to think globally and act locally, a perception that has come to increasingly determine local action. The process also involves weighing local decisions to assess their impact on the future of both the community concerned as well as that of its environment. It also entails the establishment, maintenance and further development of networks from the bottom to the top, from the local level through the sub-regional to the national level, and from there to the regional and global levels. Therefore, it is necessary for cities to network at the national, regional and global levels to learn from each other's experiences.

In a country like Germany, it is particularly important to perceive and recognize this network of links since the Constitution grants municipalities the right to regulate all local affairs for which they are responsible within the limits prescribed by law. The municipalities are also granted financial autonomy. That is to say, they have at their disposal funds that they can disburse independently while addressing the tasks and responsibilities before them. Thus, one speaks of "all-round responsibility" in local politics, thereby underscoring the wide range of tasks for which a municipality is responsible. This might lead one to rashly infer that all we need to do is to take care of our own municipal garden and remain indifferent to the problems of others. Today, no local politician worth his (or for that matter, her) salt would entertain such a short-sighted view.

In other words, the space surrounding man remains "local" – from birth right till death. Many tasks set by the individual during this time span need to be addressed. Earlier, the situation was such that the implication of a decision could be locally contained. Today, the world has been drawn closer together – a development made possible by the advance of technology. To this extent, Olof Palme's words "You are the first generation which can no more afford to make mistakes" clearly underline the importance of the global dimension.

3. Pillars of Local Self-Government in Germany

Democracy must function at the local level or else it will disappear in the long run. It is for this reason that special protection must be extended to local communities to facilities their development from the bottom to the top. Let us, therefore, examine the status of towns and municipalities within the system of government, taking Germany as example with its five pillars of local self-government:

i. Democracy at the local level is constitutionally guaranteed by the right to local self- government enshrined in *Article 28 of the German Constitution* (the *Grundgesetz* or Basic Law). The people choose their representatives through the medium of democratic elections. A more recent clause lays down that persons from other EU member states are also eligible to vote in these elections. Members of local parliaments – that is, local and town councils – are directly elected by the citizens, as are mayors, barring a few exceptions. The duties and responsibilities devolving upon these councillors are likewise set down in the Basic Law. These encompass all matters pertaining to the local community.

ii. Whoever bears responsibility for addressing local issues, as stipulated in the German Constitution, must also have the requisite *financial resources* at his disposal in order to fulfill this responsibility. A provision in the Constitution stipulates that in order to ensure local self-government, financial autonomy must also be guaranteed. In other words, to enable them to fulfill their duties, municipalities must be assured of the requisite volume of funds.

iii. What constitutes the requisite volume is an issue that has proved highly contentious at the practical level. Revenue from federal, state and municipal taxes to the tune of DM 800-900 million are distributed in a complicated procedure. The purpose of a *financial equalization scheme* between the *laenders* (federal states), and their respective municipalities, is to even out disparities in financial capacities and enable the financially weak laender to catch up with the general average in the long term. In this connection, the land of Bavaria is a good example of a once poor state that with financial assistance from the richer states invested in a future-oriented infrastructure decades ago. Today Bavaria is among those states that contribute financially to the development of weaker states.

iv. The distribution of tax revenue receives a considerable amount of attention in the German Constitution. The primary objective of this distribution is maintaining the *equality of living conditions* throughout the Federal Republic. At the municipal level, these living conditions are reflected in the provision of civic amenities and fulfilling of basic needs, such as roads, jobs, schools and other public services. By meeting this requirement embodied in the Basic Law, we have ensured that in a country like Germany with a population of 83 million – constituting the single largest population among the EU countries – the prevalence of large urban centres like Berlin, Hamburg, the Ruhr district, Frankfurt, Stuttgart or Munich has not triggered off an influx from rural areas, which would have made it impossible to maintain high standards in supplies and an efficient system of governance.

v. Another pillar enshrined in the constitution of the laender, referred to as the *connective principle*, lays down that when the federation and the laender enjoin a task on the municipalities – something the Constitution entitles them to – they are obliged to frame regulations to cover the costs. Needless to say, the matter proves much more complicated in practice; but in short, the following rules applies: "Whoever has commissioned the score must also pay for it."

4. Local Democracy and Environmental Concern: Implementation of Agenda 21

The most important thing that has a bearing on local politics today is the conclusion of the Agenda 21. Here, separate chapters are dedicated to women, children and youth, certain sections of the population and their communities, non-governmental organization and communes, along with recommendations for sustainable development. It may, however, be noted that as of today, only 12 per cent of the municipalities in Germany have decided on implementing Agenda 21. A competition has now been designed to drum up publicity for the same.

There are four indicators that appear to be of special significance for future developments – namely, ecology, economy, society/social affairs and participation. Participation addresses both questions of citizens' participation in the political and social process as well as voluntary engagement to improve the lot of those sections of the population that started off being disadvantaged. Apart from democratic and voluntary participation, another objective that is pursued here is the involvement of the citizen in the Agenda 21 process.

Participation in local political affairs is specifically provided for in German municipal laws. In order to define participation in the context of municipal development, as used here, attention can be drawn to six criteria. Mention is also made of the indicators on which these criteria are rated, as has been undertaken in a report complied by four German federal laenders.

I. A high level of voluntary participation is measured on the basis of the number of charitable institution registered per 1,000 inhabitants.

II. A high level of democratic participation is measured on the basis of the turnout for local parliamentary elections.

III. Local engagement for international justice is measured on the basis of the municipal contribution towards development assistance as a percentage of the municipal budget.

IV. Equal participation and representation of women in public life is measured on the basis of the number of women deputies in the local parliament.

V. Importance given to the living environment for children and youth is measured on the basis of local contribution towards child and youth projects as a percentage of the municipal budget.

VI. Participation in the process of sustainability is measured on the basis of the number of voluntary hours contributed per 1,000 inhabitants within the framework of the local Agenda 21.

4. Institutional Framework: Condition for Participation

There are a number of elements in the democratic system of the Federal Republic of Germany that enable direct action even if no mandate has been secured through democratic elections. In this context, the principles of equality before the law and equal rights for men and women have particular relevance. Further, the state has assumed the responsibility for obliterating any prevailing disadvantages. Accordingly, the institution of equal treatment by the local officer is enshrined in the municipal constitutions.

- To begin with, Article 28 of the Basic Law provides for the formation of a municipal assembly instead of an elected body, as is the case in Switzerland. This is particularly suited to small communities.
- A referendum is not possible at the federal level in the present-day context. Land constitutions, such as the one for North Rhine–Westphalia (Article 68), contain provisions for initiating a petition for reference as a first step. This could be aimed at by bringing in a state law, amending the law or even repealing it. The petition for a referendum must be supported by at least by one-fifth of those eligible to vote.
- If the land (state) assembly does not comply with the petition for a referendum, which the land government has to submit before the land assembly without delay, then a

referendum has to be held within a period of 10 weeks. It can only be responded to with a 'yes' or 'no'. The referendum is decided by a majority of the votes polled.

- When it comes to municipal laws, that is to the municipal constitution, it must be noted that the first and foremost duty of the municipality is to keep its residents informed on matters of importance. The options available to the authorities include, for instance, convening a citizens assembly, public hearing of citizens in district assemblies, notices in official gazettes and in the local press, as well as a leaflet campaign and the direct intimation of every individual household.
- Under certain conditions to be met (such as minimum number, stipulated time frames), citizens on their part – that is, persons living in the commune who are entitled to vote – could initiate a petition for a citizens' referendum in order to compel the municipal council to decide on a matter that is of importance to them.
- If the council fails to take a decision, the citizens can – after meeting various requirements – initiate a decision.
- Yet another option for citizens' participation is the residents' petition. It enables the participation of all the residents, including foreigners, residing legally in the community, provided they live for at least three months in the community concerned and are at least 14 years old. The aim of this option is to ensure that the council takes up an issue declared to be important, to the extent that it falls within its local jurisdiction.
- The constitution for local municipalities has made provisions for foreigners committees, especially to accommodate all foreigners. All foreigners above 16, both male and female, are eligible to vote and be elected to political office. The objective here is to enable foreigners to participate in local decision-making processes. The Foreigners' Advisory Committee can discuss all the decisions of the council in advance and participate in the deliberations by presenting their views and recommendations and raising questions.

- In addition, there is an extensive right entitling a person to turn to the council with suggestions and complaints. Every person is entitled to this right, irrespective of age and nationality. As a rule, the problem is taken up by the concerned committee.
- What is important for municipal development is the participation of citizens and competent organizations representing public concerns when procedures for drafting land development and building plans are underway. Here, the citizens may articulate their ideas and reservations even at an early stage in the planning, thereby bringing their influence to bear on the draft plan. Once the draft is complete, it is notified to the public for a month during which time it may again be re-examined.
- The citizens are likewise involved at the planning stage in the case of municipal renovation and redevelopment and residential area development schemes.
- Senior citizens committees particularly provide the elderly with an opportunity to participate in local politics.
- Finally, parliaments for children and youth provide the younger generation with an opportunity to participate in local affairs.

5. Citizens' Voluntary Participation is Indispensable

Given the magnitude of the urban problems faced by citizens, there is need for voluntary participation. In the years immediately following World War II, when Germany was razed to the ground, the local politicians, both men and women, with the most basic means at their disposal began to clear and remove the mounds of rubble from towns and cities. They were assisted hands-on by ordinary people – mainly women, for most of the men were either dead or in prison. The monumental contribution of Berlin's Trummerfrauen ("rubble women") will never be forgotten.

These men and women had a vision. Much of it remained unrealized. But reestablishment of democracy and a life of contentment in a vibrant city was an aspiration shared by all. Without voluntary participation – that is to say, participation without remuneration – this would have been impossible. For not everything can be paid for with the taxpayers' money. But it is not just the local administration and the honorary local politician who share responsibility. Every single citizen is called upon to think not just of himself but also of the world around him and· its problems. As an individual, he has numerous opportunities to contribute actively at the local level, either through citizens' action groups or societies, as also at the national or international level via NGOs such as Greenpeace. What would happen if no one was willing to take up work on an honorary basis today? Since the municipality is only in a position to assure basic amenities, many areas would remain unaddressed. For all that, the basic idea behind local participation is stunningly simple. People at the local level are better acquainted with their problems and the requisite solution than those who think and decide for them in distant seats of power. It was the Prussian Freiherr von Stein who in 1808 first convinced his king to give this policy a try. This led to the first German municipal constitution. The basic concept with all its known inadequacies then remained unchanged for almost over 200 years.

Today, an ever-increasing number of citizens are realizing their responsibilities, whether they be in politics, culture or social affairs. The fire service, the Red Cross, care of the aged, nursing, support for the poor, particularly the homeless, by providing free food (as in the case of the "Bonner Tafel", where restaurants, canteens and other large kitchens turn in their leftovers everyday) or the collection of used clothes as a small contribution towards redistribution, are all services that are simply inconceivable without the voluntary participation of citizens. Experience shows that the more direct the donor's information about how his money is being used, the greater the willingness to donate. Worldwide campaigns for donations, such as by the World Food Programme, or direct campaigns for an impoverished country in Africa require far greater efforts and more effective methods, such as TV shows with well-known artists, in order to be convincing.

6. Future Cities and Participatory Democracy

In a globalizing world where jobs are scarce, there is a keener sense of competition, both at the national and international levels, to attract companies which hold out prospects for the creation of new jobs. Thus, models for the future growth of towns/cities, along with concepts for their social, ecological and aesthetic development so as to render them into attractive locations, are truly indispensable. In some German towns, for instance, mayors, councillors and administrations have got together with the citizens of the town to conduct a highly intensive and productive dialogue on how their town should look 20 years hence and in what direction it should develop.

The issues that are vital for urban development are: locational strengths offering advantages over other towns; leveling out of social disparities; the quality of life; sustainability and democratic participation; an efficient administration functioning like a service sector company and unhampered by an authoritarian mindset; a good system of public transport assuring easy access; and a friendly manner in dealing with the citizen as clients.

What has resulted from this exchange is not just a framework that needs to be implemented. More importantly, it has meant a changed attitude on the part of citizens towards their town. They are now not just consumers of municipal services – in terms of water, electricity or gas as essentials for living – but also assume responsibility in societies, organizations or associations in the areas of sport, culture, fire services, emergency assistance, care of the sick and the aged, nursing and other fields of active participation. They constitute concrete examples of participation on a day-to-day basis.

Only an urban settlement with a functioning infrastructure and attractive recreational facilities can in the long term make a mark in a globally competitive environment. What is also important is the sparing use of soil, energy and water; that is to say, adopting a resource-conserving approach. In the long run, vast urban landscapes have no option but to decentralize their structures and act with a sense of responsibility towards themselves.

This also implies that many global problems can only be solved if local bodies are sufficiently empowered through constitutionally-backed grant of requisite levels of autonomy and financial resources necessary for future-oriented reforms. Thus, a strong municipality can vigorously address its problems. It, consequently, becomes attractive to its citizens. He who is proud of his home will do everything to keep it so. Both the individual citizen as well as local associations and organizations are more convinced about the need for involvement and participation when there is responsible leadership and where the development model for the town has been jointly formulated.

There are seven areas of local action where, true to Olof Palme's words, one can no longer afford to make fundamental mistakes. They are:

1. Construction / housing / living environment
2. Development of downtown areas
3. Development of regions around towns and cities
4. Survival of rural areas
5. Conservationist policy of land development and settlement planning
6. Social stabilization in the endangered parts of the town
7. Acceptable level of mobility and development of the local and suburban transport system

Even as local areas of action are stressed, a very prominent part will be played by global networking. Until the first half of the 19th century, Europe had not been witness to any significant change in communication since the Roman times, other than the pushcart as a medieval innovation used in road construction. Global networking was neither important nor conceivable. Road transport and communication was worse than about 2000 years ago. It was only with the coming of the railway and the steam ship in the 19th century that cities, countries and entire continents came closer together. Today, in the digital era, one need not necessarily move from place to place to be up to date. Instead, one has the option of simply accessing information from all parts of the world on one's home computer screen.

But modern communications technology has its own set of risks. The flood of information available to us must be carefully sieved so that it can appropriately handled. The GDR, which collapsed in 1990, knew everything about its citizens and the prevailing mood among them. But since it was both unable and unwilling to use this information to formulate a future-oriented policy, it disappeared into the trash bin of world history.

There is yet another danger that cannot be overlooked. Digital proximity can also lead to global alienation. Politics must counteract this. Global virtual proximity requires real-life solidarity in practical terms. One of the options that can be explored is twinning between towns to help people understand each other, learn from each other and help each other.

Yet there is reason for optimism. Education for both men and women, with technical progress, will transform living and working conditions and consequently prove to be the most effective weapon against poverty. Cities, or rather urban areas, must carry democratization forward in their decision-making process and bring local areas closer to their citizens. A firm policy of decentralization in the allocation of responsibility and financial resources is necessary for achieving the requisite structures.

In this process, top priority should be given to an urban development approach that is resource-friendly. This should be accompanied by good governance – in other words, regulation and cooperation committed to the well-being of citizen – as well as by the strengthening of civil engagement. Economic growth and participation in the development process of urban-dwellers who live in poverty are important factors in effecting improvements in the environment.

11

German Cities and Municipalities
On the Road to Sustainable Urban Development?

Helmut Dedy

I Cities, Municipalities and Sustainability

Following the Earth Summit in Rio de Janeiro in 1992 and Habitat-II in Istanbul in 1996, the objective of sustainable urban development found wide acceptance in Germany. This is true of both the national level (federal) and the local level, i.e., cities and municipalities. For both sustainable regional development, or the land use policy,[1] as well as for the transport policy,[2] the principle of municipal self-governance is of great significance. According to this principle, cities and municipalities must be guaranteed the right to regulate all the affairs of the local community for which they are responsible, within the limits stipulated by the law.[3] This principle is part of our Constitution.

In Germany, due to institutional reasons, it is not possible to realize the concept of sustainable urban development at the national level alone. Sustainability means taking the objectives of economic growth, social development and environmental care into consideration. All the three objectives also get influenced by the local municipal policy – be it positive or negative. For instance, whenever a city creates the prerequisites for the setting up of an international company, or changes the emphasis of its transport policy from road to rail, or drives out polluting production units, it directly or indirectly influences the objectives of a sustainable development policy.

Fountainhead of determinants

Sustainable development is not possible without cities and municipalities. Cities are the "fountainhead of determinants." The decisions made by them can be "right" or "wrong", depending on whether they reduce or increase environmental pollution by encouraging certain consumption patterns or permitting specific production methods. Often a city's locational decisions have had an adverse impact because they failed to consider the implications for the environment. Thus, a city can be the fountainhead of negative determinants. On the other hand, if a city adopts a sustainable transport policy, which also covers the surrounding countryside, it can become the fountainhead of positive determinants.

In Germany, 80 per cent of the 81 million people lives in cities. But these statistics can be a misleading generalization. A city in Germany can have 3.5 million inhabitants, as in the case of Berlin; and a city can also comprise merely 20,000 inhabitants. Therefore in Germany the concept of city does not always and automatically stand for urban life. If the cities of over a million inhabitants and the conurbations are taken together, they account for approximately half the German population, which means just 40 million inhabitants. Thus, Germany may be viewed as very urban or very rural depending on what concept of a city one has.

The picture is further complicated by the fact that the structures in Germany radically differ in the old laenders[4] and in the new laenders.[5] Around 15 million inhabitants live in the newly formed laenders in 6,000 municipalities, half of which accommodate less than 500 inhabitants.[6] On the other hand, 66 million inhabitants live in the old German states in 8,500 municipalities. Hence, apart from the difference between cities, or say conurbations and rural regions, another distinctions namely, between, the old and new German laenders, also adds to the urban complexity.

In the following discussion on urban development, the focus will be on big cities and conurbations.

II. Development Trends in German Cities

Notwithstanding the discourse on sustainability, three trends are discernible in Germany's urban development.

1. Settlement dispersion

Growing settlement pressures have led to dispersed and scattered urbanization. In other words, urban settlements have pushed into the surrounding regions, eating into large tracts of land. As a result, there is a sharp decline in green belts, on the one hand, and a steep rise in plastered and cemented areas and in urban sprawl around the cities, on the other.

2. *Monofunctional Units*

In societies that are based on the division of labour, there is a clear separation between residential spaces, business/ commercial areas and leisure/entertainment centres. Such monofunctional units, in turn, lead to inefficient and wasteful land use and add to the avoidable traffic congestion and rush.

3. *Growth in traffic*

The sharp increase in both personal and commercial transportation – with the number of passenger vehicles, trucks and lorries on the roads fast multiplying continues to aggravate the problems of air and noise pollution, intensive land use and degradation of natural regions.

These three trends – namely, settlement dispersion, the monofunctionality and growth in traffic – together from a vicious circle. Regional development science talk about an "upward spiral of growth in traffic and expansion of land settlement areas." The more land settlement structures become dispersed, the longer the distances to be traversed between various individual activities and greater the traffic rush. The resulting urban congestion leads to a deterioration of residential areas and an increase in migration. As a result, locations of different functions get further dispersed and are organized such that the public transport system can scarcely cope with the demand. A key question is: How can this upward spiral be checked?

III. Sustainable Urban Development: Optimizing Potential Without Robbing the Dynamics

How can urban development be rendered more sustainable? We will highlight three aspects, in particular, that are under discussion in Germany.

1. *Resource utilization in cities*

Urban development planning must change the way we use resources. First, with regard to renewable resources, their regeneration rate has to be taken into account. In other words, one shall consume only as much as is regenerated. In the context of transport, this implies that transport-generated pollution must be in proportion to the absorption capacity of the environment.

Similarly, nonrenewable resources, land in particuiar, should be used economically and carefully. Any urban area management must satisfy the criteria for sustainability. This means that the possibility of replacement must also be considered. For instance, if an area is allotted for a new local transport use, then it shall be compensated for by creating free place elsewhere. Thus, the objective of preservation of the natural capital stock must be kept in mind even in the case of nonrenewable resources.

2. *Balancing material exchange between cities and the surrounding countryside*

Apart from sustainable resource use within cities the exchange processes between a city and its surrounding countryside also require special attention from a regional perspective. The city and its surrounding countryside exchange workforce, intermediate products and machines, goods and services, waste and water and so on. Therefore, it may be naïve to declare these processes simply as ecologically harmful. It is not a question of prescribing "autocracy" to certain extent, for the cities, even if priority is given to the objective of "sustainability". A city just cannot exist without its nearby suburbs and farther countryside – a functional regional division of labour will always be necessary. Therefore, a complete withdrawal of processes based on the division of labour is not recommended. Instead, an attempt should be made to render the exchange processes between the city and the surrounding countryside ecologically sustainable.

Let us take the export of particular waste as an example. It is surely in favour of Hamburg to dispose of its waste at a site situated more than 500 kilometers away at some other place in Germany. The waste is gone, and Hamburg remains pollution-free. However, for the location of the waste disposal site this decision is linked to varied types of pollution. Moreover, for the regions situated en route between Hamburg and the disposal site, it means considerable amount of traffic-generated pollution. This is one view. According to another point of view, Hamburg will always be dependent on a disposal site, situated anywhere in Germany; therefore, (almost) any desired price shall be paid. And at the location of disposal site, earnings, jobs and economic opportunities are created. Thus, there is a mutual carrying and easing of burdens for Hamburg city and the disposal sites – only the regions through which the traffic follows have to reckon with pollution. Clearly, it is important to view issues from a broader perspective, i.e., in terms of regions and, in this respect, to check the egoism of cities.

In a society based on the division of labour another problem of material exchange arises due to different cost structures. It may, for instance, make sense in terms of cost and business management for a company in Europe to transport potatoes more than 1,000 kilometers away for peeling; to transport the peeled potatoes back to the original place so as to process them into potato chips, which then are consumed at the place situated 1,000 kilometers away from where the potatoes were peeled. This may be very clever from the point of view of business management. However, as far as the national economy and usage of available resources are concerned, this procedure is completely absurd. Thus, an expensive urban development process can mould the material exchange between a city and its suburbs; but such processes also partly stem from the society that is based on the division of labour. It opens up new areas of political action and, at the same time, reveals the limitation of a sustainable urban policy.

3. ***Sustainable urban development calls for a new regional order***

The objective of the discourse on sustainable urban development is a new regional order. It should be characterized by the following features:

- More compact , i.e., should prevent unnecessary expansion of cities into surrounding areas and hence regulate the land use

- Mixed land use, i.e., should enable working and living in the same municipal district and hence avoid unnecessary commuting;
- More concentrated within the framework of decentralization; i.e., should ensure that development takes place specifically at selected places and that cities do not develop in a dispersed manner

IV. Current Areas of Action for Sustainable Urban Development

How will sustainable urban development in Germany be put in to practice by city planners? At present a number of measures are being debated. They include a regional development plan, urban development and a transport policy. Some of these issues are discussed below.

1. City with short routes?

The bigger the city or conurbation, the stronger is the tendency towards dispersion and land use in the suburbs. The leisure park lies in the west, service providers work in the center, where scarcely anyone lives, and the industry is situated in the south. Hence German cities are increasingly striving for a combination of functions, namely, living, support services, working and leisure. However, "city with short routes" is perhaps realistic only when it also has usable waste areas situated in attractive locations. Not only the place vacated due to the possible change from industrial country to a service-providing country,[7] but the withdrawal of the troops of the Allied powers after the reunification of Germany in 1990 also serve this purpose. Many large areas of industrial units, former military installations or former railway places, which are often centrally situated, can be used for city reorganization. Therefore, a conception of urban development premised on existing resources, and the development and reutilization of waste areas in the centre of the city, can make a city with short routes a reality.

2. Are poor and rich coexisting peacefully?

In Germany, the gap between "rich" and "poor" municipal districts is rather wide. As a result, different social strata have little overlap. Therefore, the cities are not only in need of a basic reserve of state-subsidized apartments but also of the opportunity to provide middle-income groups suitable schemes for home ownership. However, well-balanced social integration is not the only item on the agenda. The recent liberalization of the country's immigration policy has raised new issues, such as the integration of different cultures and mentalities. This might be one of the biggest challenges for German politics.

3. *Creating broader terms of reference for planning*

With development thinking along regional lines acquiring increasing significance the identification and earmarking of areas for house construction and industrial development should be better coordinated between cities and municipalities in the regions. In congested areas, often the hub of the city accommodates the majority of workplaces, whereas the residential areas are pushed out to the suburbs.

The endeavor then is to think in terms of a larger context and to create instruments – say, in the environs of Hamburg and Munich – that encourage cities and municipalities, to adopt coordinated land-use policies. Perhaps the expression "controlled competition" comes closest to this idea. Competition between cities and municipalities is desirable; and here Germany is lagging behind. However, the competition for industrial development must rest on structures that ensure that the region as such remains worth living in.

4. *Land settlement and transport planning are closely linked*

A policy on housing, urban development and land settlement, which strengthens the regional development thrust, can only be successful if it integrates local and regional transport planning and transport policy. Consequently, cities, conurbations in particular, must match and integrate their settlement planning and transport planning. Extension and reconstruction of housing estates should be planned only along efficient routes of public transport; companies too require this infrastructure, as also efficient connections for goods transport.

5. *The region pushes back the city*

The aforementioned points make it clear that ultimately it is a question of strengthening cooperation within a region. Competition at the local level in the domestic market of European Union has intensified this debate. All of a sudden, individual cities or municipalities no more compete with each other. Instead, it is the regions that are competing – developing a common identity and increasingly comprehending their respective area as a common planning region. The new concept of development is underpinned by this regional thinking. Patterned after the European development model, planning is conceptualized within a regional framework rather than in local terms.

V. Sustainable Urban Development Gives Rise to Conflicts

1. *Conflicts must be defined*

The path of sustainable urban development is fraught with conflicts. For instance, considering the prices of plots in agglomeration, the objective of waste disposal near town at a reasonable cost is not always feasible.[8] Also, the objective of a compact regional order does not always correspond with the wishes of many people, who would certainly like to stay near the city but in a house with a garden and greenery. Such conflicts must be clearly defined right at the outset, so as to avoid jeopardizing the objective of sustainable urban development. It makes no sense at all to downplay the difficulties or incompatibilities for individual companies or private households – as soon as the planning process is concretized, these conflicts surface anyway.

There can be no sustainable development without the involvement of concerned parties, nor can it come about against their wishes. Unless the development path adopted has a broader basis – through cooperative and flexible planning and decision-making processes, with partners having equal rights and powers – a radical structural change is bound to remain a vision. Only the active involvement of various agencies may help to clarify the social and economic conditions under which individuals actually favour sustainable development, rather than merely appreciating and accepting it; and the conditions under which are they willing to effect changes in the production processes and their lifestyles.

2. *Involving people in planning*

People concerned must come forward to support and contribute to the process of sustainable development. Ordered and administratively enforced changes in conduct generally generate losses due to resistance but yield no sustained insights. In addition to meeting the official performance targets, a willingness to voluntarily support the objectives of sustainability within the scope of an individual's responsibility is also required. For this purpose, a critical mass of people must be convinced of the necessity for this strategy.

Support for the objectives increases in proportion to the level of participation in the process of setting and executing the performance targets. The latter, in turn, is influenced by the degree of openness and transparency that informs the process. An effective communication strategy is vital for the successful adoption of a sustainable development model.

VI. Conclusion

In Germany, an urban development policy can become sustainable only if it increasingly acquires the status of a policy for regional development. At present, a number of measures are under discussion, such as regional development planing, urban development and transport planning. Thus, it is imperative that the individual objectives of a sustainable policy be brought together by taking recourse to planning in broader interurban contexts. This is going to be the main challenge in the years to come for cities and municipalities that have the right to regulate all the affairs of the local community for which they are responsible, within the limits set by the law.

Appendix I

Some Details of the Regional Development Plan and Land-use Policy in Germany

I. Constitutional and institutional framework

Federally structured Germany has an elaborate administrative set-up. The federative principle in Germany leads to the intertwining and interlocking of responsibilities and financial relations of the federation, laenders and municipalities. The same applies to the political field of regional and urban development. The distribution of tasks among the federation, laenders and municipalities represents the subsidiarity principle of Germany's federal system. Each of these units have their respective sets of authorities and responsibilities for regional and urban development.

- *Federation:* The federal government, ministries and subordinate authorities are empowered to provide skeletal legislation for regional and urban development. The Regional Development Act (ROG) and the Town and Country Planning Code (BauGB) form the legal basis of regional development. The responsibility for formulating land settlement policies, i.e., urban and regional development, lies with the ministries of transport, construction and housing.

- *Laender:* Each federal state is responsible for enforcing not only its own laws but also a major portion of the federal laws. Each laender has a ministry that concerns itself with issues related to regional and urban development. It regulates urban and regional development within the scope of its competence. Laenders control the extensive development trends with the help of country planning. Additionally, they make their own urban development policies appropriate to the prevailing scenario and requirements, within the limits defined by ROG and BauGB.

- *Town and municipalities:* These local political authorities have the right to self-government. Municipalities are guaranteed the right to "regulate on their own responsibility all the affairs of the local community" by the Basic Law (Article 28 para 2). The "institutional guarantee of local self-government", however, is subject to the provision that it shall be exercised "within the limits of the law (of the Federation and Laender)." One of the important elements of local self-government is the right to take over the responsibility of planning in the field of urban development and land use within the local boundaries and as per the limits set by the BauGB (planning authority).

2. *Objectives, strategies and priorities at the national, regional and local levels*

The Regional Development Act (ROG) and the Town and Country Planning Code (BauGB) define the "main objectives" of regional and urban development policy. The principle of sustainability is anchored in Section 1 of ROG. In the principles of regional development (Section 2), it is mentioned, among other things, that a decentralized land settlement structure shall be maintained, land settlement activities shall be concentrated regionally and reuse of unproductive land settlement areas instead of open space shall be given priority. Further, BauGB stipulates an economical and careful handling of land and soil (Section 1). With the Guidelines (ORA) for regional development policy, an attempt was made to realize these main objectives and transform them into models for regional development. The ORA guidelines, which were adopted at the ministerial conference for regional development (by a federation-laender board for coordination of regional development policy), are viewed as models for regional development. The ORA's model for environment and land use give priority to, among other things, the objectives of housing estate development:

- In the environs of congested and compact areas, maintenance and improvement of (remaining) open spaces is given top priority in order to ensure and firmly establish the workability of spatial structures in a regional context.
- A preventive policy is required for minimizing conflicts related to land use. Existing encumbrances form the basis of new measures and are taken adequately into account. For reducing land use conflicts, infrastructure-related facilities must be largely correlated and coordinated.
- Conservation of regional resources and environment begins with a land settlement structure and land use. Therefore, a meaningful allocation of workplace and housing locations, as well as an appropriate allotment of land on a regional scale, contributes to the avoidance of additional environmental pollution.
- Individual regions must see to it that the environment is and remains an important locational factor. Environmental potential is to be exploited "economically" – with the aim to retain the natural regeneration capacity of the land. In this respect, each region takes the responsibility for shaping its own ecosystem.
- A sensitive and intelligent development of decentralized areas and land settlement structures safeguards the environment. This calls for a far-sighted land-use policy, especially in conurbations.

Appendix II

Some Details of the Transport Policy in Germany

II. Constitutional and institutional policy framework

As in the case of regional and urban development, there exists in the sphere of transport planning a close interrelation between the federations, laender and municipalities. The federation promotes plans for improving transport facilities in the municipalities as per the Municipality Transport Financing Law (GVFG). This applies both to projects of local public transport (OPNV) and for local road construction. For investment towards improving the OPNV alone, the federation provided over 50 billion DM to the laenders and Deutschen Bahn AG between 1967 and 1998. These funds have been predominantly earmarked for agglomerations. Items to be promoted included, for instance, urban underground and suburban railway, parks with rides, central coach stations, interchange facilities, bus lanes and cycle-ways.

The reformed Road Traffic Regulation (StVO) provides municipalities with a set of policies that gives them room for manoeuvre at the time of solving traffic-related problems. This covers, among other things, the introduction of speech limit zones regulation (Tempo 30) or the identification of silent zones.

2. Objectives, strategies and main issues at the national, regional and local levels

It is an undisputed fact that a city good for automobiles — with its attendant problems of environmental pollution, through noise and exhaust, accidents, and shortage of parking places, is not worth living in. The same, applies to a city without automobiles. It is, therefore, necessary to find a balance between the mobility demands of the population and the economy and measures that ensure quality of life. These issues need to be addressed particularly in agglomerations. For this purpose, the following measures are necessary:

- OPNV should be made attractive and efficient.

- Short-haul transportation should be included in the integrated state transport concepts.
- Every agglomeration should have its own structure. Therefore, appropriate solutions must be found that fulfill the respective prerequisites.
- The networking of public carriers in passenger and goods transportation should be improved.

For the implementation of these objectives, a distinct approach and definite instruments are available. It is of the utmost importance to find local solutions to various problems. There are no general rules applicable for the whole of Germany.

NOTES AND REFERENCES

1 Refer the text in Appendix 1 on constitutional and institutional policy framework for land use.

2 Refer the text in Appendix 2 on constitutional and institutional policy framework for transport policy.

3 Article 28 Para 2 of the Basic Law

4 It refers to the states in Federal Republic of Germany prior to the German reunification in 1990.

5 It refers to the states in German Democratic Republic prior to the German reunification in 1990.

6 However, it must be pointed out that the new Laenders have experienced problems because of their small size. At present, they are carrying out municipal area reforms, which should lead to larger municipal units.

7 Of course, such developments vary from region to region; for instance, the reorganization of the former coal and steel region of north Rhine-Westphalia.

8 This applies so long as the external costs are ignored – in case of Hamburg, the traffic–related pollution in the region.

PART II

Case Studies

12

Pathways to Responsive and Sustainable Environmental Improvement for the Urban Poor

The Example of Andhra Pradesh

Banashree Banerjee

1. Well-Known Problem

One of the reasons for the mounting disillusionment of poor urban communities with municipalities is to do with the latter's inability to bring about adequate improvements in the living conditions of the communities. This situation obtains in spite of the large number of government and donor-funded programmes underway for slum improvement and the substantial investment over the years in laying water pipes and building latrines, roads, drains and garbage dumps. Evidence shows that investment in engineering works does not necessarily guarantee environmental improvement.

There is, of course, the question of numbers: the poor outnumber the efforts of local governments to extend civic services. The assistance provided by the state and central governments and donor agencies is also inadequate. To cover as many people as possible, the general practice is to spread the resources thinly with little impact. The other practice is to reduce the initial costs by lowering standards and specifications. The near absence of maintenance ensures rapid deterioration of these substandard services.

Local residents are often expected to undertake operation and maintenance tasks. However, this seldom happens because their needs, practices and priorities are not considered in the 'technocratic' approach to infrastructure improvement. Indian cities are littered with abandoned latrines and choked drains.

Another reason for the unresponsive and unsustainable services in poor neighbourhoods is the narrow view of environmental improvement. First, in-settlement services are often not linked with city services. This is particularly problematic in the case of drains, where improvements can sometimes lead to a set of new environmental problems. Second, hardly any attention is paid to wider environmental problems, such as flooding, water stagnation and landslides. These problems arise from the marginal location of poor settlements in the city. The focus on the in-settlement upgrading of services can do nothing to improve the situation.

What is the way out?

Obviously there are no easy or universal answers. But there is no doubt about the centrality of the role of the municipality in making services accessible to the poor. Looking at the issues underlying the manifestations of the problem, it is clear that the answers lie in changing the way municipalities function in four important ways:

1. Building their capability to dedicate or mobilize funds for improving the environment of poor neighbourhoods
2. Demonstrating the political willingness to commit resources for operation and maintenance
3. Instituting participatory processes that legitimize demand making by civil society, particularly the poor, and that enable municipalities to respond to those demands
4. Undertaking environmental improvement in the city-wide context of service networks and environmental problems

Fundamentally, this implies a process of municipal reform, comprising better ways of defining problems, planning and finding solutions acceptable to the target communities.

The seeds of such a reform process already exist in the form of the 74th Constitutional Amendment Act (74th CAA) of 1992 for decentralizing power and responsibilities to urban local bodies(ULBs). The specific responsibilities in relation to infrastructure improvement for the poor are: (i) safeguarding the interests of the weaker sections of society, including the physically handicapped and mentally retarded, (ii) slum improvement and upgradation, (iii) urban poverty alleviation and (iv) urban planning (Ministry of Urban Development 1992).

The other starting point could be the municipal experience of working with community groups in prioritizing local services; supporting community contracting of works as part of the NSDP (National Slum Development Programme); and, to a limited extent, the experience of working with NGOs (non-governmental organizations) in addressing local priorities.

There could be different ways of undertaking the required reforms. One step forward has been shown by the Andhra Pradesh Urban Services for the Poor (APUSP) project currently being implemented in Andhra Pradesh (AP) state in southern India. This paper examines the principles and processes by which the project has been building a municipal reform agenda towards responsive and sustainable infrastructure improvement for the poor.

2.0 Introduction to APUSP: Three-pronged Approach

APUSP covers 32 Class 1 towns[1], which have a combined population of 5.9 million (Census 2001). The project goal is, "sustained reduction in poverty and vulnerability of the urban poor in Andhra Pradesh." The purpose is that 'the poor in Class 1 towns of AP benefit from improved access to more appropriate and sustainable services (DFID 1999)."

In its seven-year period, the project is expected to directly benefit more than 2.4 million slum-dwellers[2] in the 32 towns. However, in a large country, which has been undertaking government and donor-funded poverty reduction programmes for the last three decades, it is neither the scale nor the goal for which APUSP deserves attention. The merit of APUSP lies in its approach to poverty reduction.

The concept of poverty reduction in APUSP is based on three premises. First, reforms are required in municipalities in order to improve their performance in poverty reduction activities. Second, improvement of environmental infrastructure in slums has multiple implications for improving the conditions of the poor, but it should be based on the demands of the poor and on the capacity of the municipality to operate and maintain it. Third, civil society needs to be strengthened as a way to invigorate the interaction between the poor and the municipality and to stimulate policies in favor of the poor. This reflects the notion that poverty reduction requires the building of partnerships in both government and community or civil society, and that it needs to be based on the felt needs of the poor.

This concept has been translated into a three-pronged approach, consisting of funding for three inter-linked components: component 1: municipal reform; component 2: environmental infrastructure improvement; and component 3: working with civil society. The three components are expected to mutually strengthen each other in a number of ways (DFID 1999). (See Figure. 1).

The first two components are being undertaken by the state government. They have been brought together in a medium-term rolling plan, the Municipal Action Plan for Poverty Reduction (MAPP), articulated by each of the 32 municipalities, with the participation of local stakeholders. Over time, the parallel civil society initiative is expected to improve the participatory quality and the poverty and vulnerability focus of the MAPP. It is also expected to influence state policy.

Figure. 1: Three Linked Components

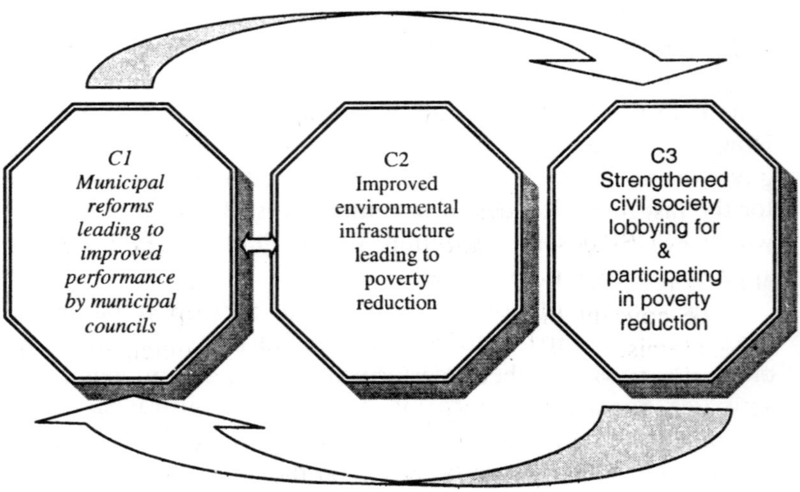

Source: Based on Basic MAPP Guidelines (APUSP 2001).

The first year of project implementation (2000-2001) has demonstrated the validity of the approach. It has also brought into sharp focus the complexity of putting into practice some of the key principles of APUSP. In the following sections we will look at its practical implications. But first, we will review the project components and the MAPP process in order to arrive at a better understanding of their contribution to responsive and sustainable environmental improvement for the poor.

3.0 Contribution of Project Components to Responsive and Sustainable Environmental Improvement for the Poor

3.1 Municipal reforms

Municipal reforms in APUSP aim to improve the manner in which municipalities undertake the responsibility for poverty reduction. This involves a wide range of improvement measures. Municipalities need to set up consultation mechanisms with the poor in order to identify and respond to their priorities. They need to mobilize adequate skills and finances to provide services as well as operate and maintain them on a long-term basis. Even though APUSP does not provide funding to municipalities for the non-infrastructure needs of the poor, the project expects them to coordinate poverty reduction activities in such a way that these needs are not overlooked.

Municipalities have the legal obligation to undertake poverty alleviation and slum improvement under the 74th CAA. But so far, decentralization has meant additional responsibility, without additional powers or resources, in weak systems. Municipalities largely depend on central or state government funds for carrying out their activities. These funds are not always assured, lending a high degree of uncertainty to municipal functioning. Planning of any sort becomes redundant and is therefore neglected.

However, a change for the better is expected because of the initiatives of the Government of AP in the last five years. The thrust is on good governance and financial discipline for affording greater autonomy to local bodies. Measures such as incentives for higher revenue collection, support for computerization and technical assistance for updating accounts have resulted in the improvement of the revenue position of most municipalities. Introduction of user charges and private sector participation in service delivery have been tried on a small scale in a few municipalities. Greater accountability in routine municipal functions has been introduced, and single window citizens' service centres have been established for the convenience of the people of the town.

Parallel to decentralization has been a process of democratization and grassroots representation. The 74th CAA recognizes local governments as the third tier of government. Municipal councils are democratically elected, with one-third of the council seats being reserved for women.

In addition, the central government's poverty alleviation schemes require municipalities to facilitate the formation of bodies that have representatives of women-only community-based organizations in slums. Resident community volunteers (RCVs) are elected for a neighbourhood group (NHG) of 20 to 40 households; RCVs in a slum or ward[3] form a neighbourhood committee (NHC); one of the RCVs is elected as a convenor. NHC convenors form the community development society (CDS) at the town level and elect their own office bearers (Ministry of Urban Affairs and Employment 1998a). Programmes for the economic, social and physical development of the poor are expected to be implemented by the municipality with the active involvement of the CDS structure. This has institutionalized the representation of poor women and given municipalities some experience of working with them.

The decentralization and democratization initiatives of the government provide a good starting point for reforms to sustain poverty reduction. The strategy adopted by APUSP is to build on the existing initiatives of good governance and financial management on the one hand, and strengthen the participatory structures of the poor, on the other. The basic tasks are to support municipalities to bring about targeted improvements in financial management, planning and implementation capacity, and mechanisms for consultations with citizens, particularly the poor. The broad indicative areas of reform are: revenue improvement, institutional development, participatory planning approaches, service delivery improvement, operation and maintenance, training/skill upgrading and other matters appropriate to a town's circumstances.

The premises of the targeted approach to reform are as follows: Better ways of interfacing with citizens and their involvement in making decisions that affect them will improve targeting. Improved revenue assessment and collection will result in the generation of resources. Improved management of funds, through transparent accounting and budgeting systems, will lead to the allocation of resources for the provision of services for the poor and for their operation and maintenance. This, in turn, will lead to better user satisfaction. It will provide the incentive for municipalities to take measures such as cost recovery and increase in taxes, which may otherwise be politically difficult. This process will be further stimulated by a more vocal and informed civil society, which will impress on the municipalities the need to improve services for poor communities and involve them in decision-making processes.

The central issue is that improvement in performance requires willingness on the part of municipalities to change. Reform, driven by government policy and state government directives, is one way of bringing about change. In the longer term, the pressure created by a strengthened civil society may also induce change in municipal working. However, if reforms have to substantially benefit the poor, the incentives have to be different.

In APUSP, funding for infrastructure improvement for the poor is expected to provide the incentive to reform. Measurable performance improvement is the trigger for gaining access to APUSP project funds for infrastructure improvement. There are "… no automatic entitlements for those municipalities who do not demonstrate a commitment to change, to address the needs of the poor and the ability to operate and maintain the infrastructure provided (DFID 1999)."

It is worth mentioning that the incentive system for pro-poor planning is breaking new ground in AP and in India. Such a system is very different from the present policy and practice of equal shares for different administrative units (district, town and ward) based on per capita allocations. Allocations may be topped up for particularly vulnerable groups, but they are never based on performance.

3.2 Environmental infrastructure

The focus of APUSP on environmental infrastructure improvement is in response to the current infrastructure deficiency and the inability of local government to deal with the situation. Improvement of environmental infrastructure for the poor is urgently needed in most of the 32 towns covered under APUSP. This is well-illustrated by the studies carried out during the design of APUSP (See Table.1).

Table. 1: Environmental Infrastructure Deficiency in Slums in Class 1 Towns of AP

Infrastructure Deficiency	Per cent slum households
No municipal water supply	46
No latrines (open defecation)	29
No all-weather roads	41
No drainage	45
No street lights	20

Source: APUSP project design studies (1998), DFID.

It is a well-known fact that municipalities do not generate sufficient revenues to cater to the service needs of the growing population of poor people. This deficiency has been taken care of to some extent by several central and state government schemes that have been implemented during the last three decades. However, their impact has not been significant. The allocations have been meagre compared with the large and growing need and have been thinly spread for the sake of sharing the benefits across political constituencies. The quality of the services is invariably poor, and deteriorates rapidly with no attention to allocations for operation and maintenance. The schemes were top-down, with no scope for municipalities or poor communities to have a say in the way they are implemented. A significant change in the new programmes being implemented since 1998 is the involvement of women-only representatives of community-based organizations in prioritizing infrastructure needs in slums and in implementing some of these programmes through community contracting

(Ministry of Urban Affairs and Employment 1998 a and b, Rao 2000).

APUSP project funds are primarily for in-settlement upgrading of water supply, roads, drains, street lights, latrines and garbage disposal; and also for the infrastructure necessary to link internal services to town networks. The adverse effects of environmental problems such as area-wide flooding are also taken into consideration. This is different from previous slum improvement programmes, which did not consider linking arrangements and wider environmental problems, often leading to drains without out-falls, taps without water and persistent flooding. Hygiene promotion is included in APUSP as an adjunct to infrastructure improvement to enhance the health benefits of improved services.

Infrastructure as a means to poverty reduction is justified because infrastructure improvement has a wide-ranging impact on poverty. This has been amply demonstrated in an impact assessment study of DFID-supported slum improvement projects in India. The study pointed out that though infrastructure contributes primarily to improving the quality of life, it also has a positive impact on the survival and security dimensions of poverty by reducing the vulnerability to disease and improving the conditions for income enhancement (Amis 2001). The lessons from this study have informed the design of APUSP. These findings are also supported by the perceptions of poor communities in the Class 1 towns of AP. People living in slums, who participated in appraisal exercises during the design of APUSP, identified multiple benefits of water supply, roads, drains, latrines and electricity.

The stress on municipalities undertaking infrastructure improvement does not mean that lack of infrastructure is the only deprivation faced by the poor in Class 1 towns. Instead, the implication is that project municipalities have some capacity to address the infrastructure needs of the poor, but next to no capacity to address the more difficult problems related to social deprivations, employment and entitlements. Even the better-equipped municipal corporations in the large cities where the DFID slum improvement programmes were carried out could not do justice to the social and economic components of those projects (Amis 2001). The alternative strategy in APUSP is to address non-infrastructure needs through the civil society component.

3.3 Strengthening civil society

The implication from the previous sections is that strengthening civil society represents an attempt to strengthen the demand-making capacity of citizens, particularly poor communities, towards the municipality. This is a very important part of the civil society component. The other part is related to an attempt to find alternative channels, through civil society organizations (CSOs),[4] to address the wider survival, security and livelihood problems faced by the poor.

This component of APUSP is a parallel stream because the partners, channels of assistance and types of activities differ from the governance and service provision focus of the other two components. Independent funding and technical assistance is expected to provide an incentive for establishing links with other non-governmental networks and funding sources, widening the ambit of poverty-reduction activities.

The civil society initiative in a town starts with a participatory poverty assessment (PPA), which involves different stakeholders in the town. The PPAs explore local concepts of poverty and vulnerability, livelihood strategies of the poor, people's perceived priorities, local views on programme/policy implementation, etc. Issues highlighted in the five PPAs carried out so far include aspects such as insecurity of tenure and employment, alcoholism, violence, caste discrimination, difficulties in getting access to institutions and entitlements, low-qualify services, poor health, vulnerable groups such as deserted women, elderly, unemployed youth, etc. The issues are not the same in different towns and or even in different neighbourhoods of the same town.

The PPAs are expected to build a shared understanding of the local dimensions of poverty and to generate partnerships and alliances in civil society (including the municipality) to take up local poverty issues. The funding of projects is through the Urban Initiative Fund (UIF) which will provide windows of opportunity to stakeholders who are best suited for specific poverty-reduction activities to undertake these activities. Further, the activities will be defined in a transparent manner, based on a shared understanding of local poverty. This approach is expected to overcome the limitations of earlier DFID slum improvement programmes, where the social, health and economic support programmes were implemented uniformly across slums by municipalities, with indifferent results (Amis 2001).

The PPA and accompanying research on different findings of the PPA will also inform social policy and may have an impact on the way poverty programmes are structured. The immediate impact will be on improving the poverty focus of MAPP.

4.0 MAPP as a Framework for Responsive and Sustainable Infrastructure Improvement for the Poor

4.1 MAPP objectives and concept

Each of the 32 municipalities is required to prepare a MAPP to enable it to access APUSP funds. The MAPP enables a municipality to define how it will achieve three linked objectives in its own context. These objectives, which derive from the principles of APUSP, are:

- To improve municipal governance and management so that municipalities become more efficient and responsive to the needs of the people, especially the poor
- To improve environmental infrastructure for the poor in a sustainable way
- To identify and undertake poverty-reduction measures with the active participation of the poor and civil society

The MAPP is a rolling plan, which is updated periodically. It begins with a Basic MAPP, in keeping with the present limited capacity of municipalities, and later evolves in to a Full MAPP (See Table. 2). The MAPP contains the municipality's strategy, proposals and implementation plan, with measurable outcomes for municipal reform and environmental infrastructure, as well as the approach to the convergence of other poverty alleviation programmes.

In the first year of project implementation, the Basic MAPP guidelines were drafted and piloted, through a closely guided process, in three towns (Chittoor, Qutbullapur and Rajahmundry) and finalized after extensive consultation with municipalities. The preparation of guidelines was undertaken by the APUSP Team (municipal strengthening unit (MSU), appraisal and monitoring unit (AMU)[5] and consultants).

Table 2: MAPP Concept

Elements	Basic MAPP	Full MAPP
Time frame	1-2 years	3-5 years rolling plan
Database	Uses existing data	Develops database
Municipal reform and performance improvement	➢ Simple and few reforms ➢ Essential training for MAPP	➢ More complex reforms and improvement measures ➢ Training also for performance improvement
Environmental infrastructure	In a few wards on the basis of selection criteria	Phased yearly programme including a large no. of wards
Convergence	Only of urban poverty programmes implemented through the municipality	Includes programmes of other agencies and UIF under C3
Resource mobilization	Planned within DFID and available resources	Proposes to mobilize additional funding

Source: APUSP Project Document (1999), DFID.

The APUSP Team conducted interactive regional workshops for five to six municipalities at a time. The main objectives were to familiarize senior elected and executive functionaries of the municipalities with the guidelines and processes, and to build their capacity to undertake the preparation of their MAPPs. By November 2001, eight municipalities had completed their MAPPs and 15 were in various stages of MAPP preparation.

According to the guidelines, MAPP preparation broadly follows a nine-stage process in a municipality (APUSP 2001). It is worth summarizing the process as it contributes to developing a participatory, inclusive and transparent approach to poverty reduction in a town-wide context and within the framework of municipal planning and budgeting.

4.2 Stages in MAPP preparation

Preparatory steps: Municipalities are required to form two local committees before embarking on the MAPP process. The municipal reforms committee (MRC) is constituted to give strategic guidance to the MAPP process. It is headed by the mayor/municipal chairperson and includes the local MLA/MP, the floor leader of the opposition party and other members of the municipal council, municipal commissioner[6], two persons from outside the council and the CDS president. The municipal task force (MTF) has been given the day-to-day responsibility for MAPP preparation. Headed by the commissioner, the MTF is a multidisciplinary team consisting of heads of the functional sections of the municipality.

Stage 1: Getting started: The APUSP project and the purpose and process of the MAPP are introduced by the municipality at a large meeting to which representatives of a wide range of stakeholders are invited. The members of the MRC and MTF are introduced.

Stage 2: Problem identification: This stage starts with a workshop coordinated by the municipality with representatives of stakeholder groups (elected municipal councillors, municipal staff, CDS and other CBOs, NGOs and other organizations working with the poor). One-third of the participants are women. The participants work in groups to articulate a vision statement for the town and identify problems with respect to poverty, infrastructure and municipal performance. The workshop identifies two working groups, consisting of members representing the stakeholder groups, to provide a more detailed analysis of the problems over a period of three to four weeks.

Working Group 1 on Municipal Performance Improvement carries out an analysis of municipal strengths, weaknesses, opportunities and threats (SWOT) in relation to financial status, staffing, service delivery and interfacing with citizens. The problems and their causes are analyzed with the involvement of the entire cross-section of municipal staff and elected representatives to identify strategic objectives for municipal performance improvement, which are then ratified by the MRC. Proposals are prepared to address these objectives and assessed against criteria such as benefits, political and social acceptability, technical/managerial feasibility, cost and measurable outcomes within the MAPP period.

Working Group 2 on Poverty and Infrastructure Deficiency identifies different kinds of poor settlements and uses the existing data to rank them in terms of poverty and infrastructure deficiency and shows the ranking on a matrix. It also places poor settlements in a citywide perspective of ward boundaries, service networks and environmental problems such as flooding, sea and river erosion and pollution. This analysis is shown on maps and charts. A technical screening group to check site conditions and identify the need for off-site infrastructure visits the poorest and most deficient settlements on the matrix. An environmental statement for the town is also produced.

The two working groups meet to ensure coordination and compatibility of outputs. The fiindings of Group 2 related to the need for data improvement, network design, staff capacity building, equipment, etc., are incorporated into the municipal improvement proposals of Group 1.

Stage 3: Strategy formulation: A two-day workshop is convened with the participants of the problem analysis workshop. The findings of the two working groups are presented after which the workshop recommends which poor settlements are priority settlements for inclusion, which strategy is to be followed for infrastructure investments and which reform proposals are to be taken up in the MAPP. The MRC deliberates on the recommendations and endorses them, with or without changes. The decision of the MRC and the next steps in MAPP preparation are shared with the workshop participants.

Stage 4: Preparation of reform proposals: The MTF reviews the reform proposals recommended by the strategy formulation workshop and consolidates them into a set of proposals to meet the strategic objectives for municipal performance improvement. The expected outcomes of each proposal are considered and an implementation plan is prepared for the proposals.

Stage 5: Preparation of environmental infrastructure proposals: The first step at this stage is micro-planning with community groups in the priority poor settlements. The process is undertaken by a community micro-planning group, coordinated by the NHC convenor and consisting of all the RCVs in the settlement, representatives of other women and youth groups, field workers of programmes (literacy, health, woman and child development etc.) in the area and municipal sanitary workers. The exercise is facilitated by a multidisciplinary municipal micro-planning team and takes four to five weeks. The broad steps in the process are:

- Inventory of existing infrastructure and needs identification by the group, using participatory mapping techniques
- Preparation of a draft settlement plan for infrastructure by the group, facilitated by the team
- Technical assessment of proposals by the municipal engineering staff
- Feedback to the community and agreement on a final settlement plan for infrastructure, showing rehabilitation proposals for existing infrastructure and additional infrastructure proposals
- Discussion on different implementation and operation and maintenance responsibilities

The second step is the preparation of environmental infrastructure proposals based on the settlement plans for infrastructure and off-site infrastructure needs. The engineering section of the municipality prepares these with assistance, if required, from the local office of the AP State Public Health Engineering Department (PHED). The process is overseen by the MTF. An implementation plan is prepared for the proposals.

Stage 6: Finalizing the MAPP document: The MTF is responsible for this task. The document outlines the process of MAPP preparation and submits proposals for reform and infrastructure improvement, as detailed in stages 4 and 5. The MTF cross checks the document against the APUSP appraisal criteria. The document is presented to the municipal council for adoption.

Stage 7: Appraisal and approval: The AMU carries out concurrent field appraisal of the MAPP process and final desk appraisal according to the agreed criteria. Poverty and gender focus, participation and municipal capacity to engage in these aspects form part of the appraisal criteria. The empowered committee endorses the proposals on the basis of the appraisal report.

Stage 8: Preparation of detailed proposals: Detailed proposals for reform and infrastructure improvement are prepared after the MAPP is endorsed. The proposals are expected to clearly indicate measurable outputs at milestones, procurement and operation arrangements, and detailed costs, including operation and maintenance. The infrastructure proposals are broken up into subprojects for appropriate contract packaging. Small subprojects for roads and drains may be contracted to willing community groups. In any case, community groups are expected to monitor contracts in their respective areas.

Stage 9: Sub-project review: The subprojects are reviewed for technical and managerial feasibility by the AMU, after which funds are released.

4.3 The Basic MAPP process in practice

Experience from managing the MAPP process in the three pilot municipalities shows that the process is appreciated by the municipalities for its clarity, transparency and acceptability. There is representation of the civil society, including poor communities, at all the stages where problems and needs are assessed and decisions are made. The guidelines represent a highly participatory process, both within and beyond the municipality, which is completely different from previous practice. It is for this reason that most municipalities require substantial guidance for preparing the first MAPP. In the following section, we will review how some critical issues related to environmental improvement for the poor are dealt with in practice.

5.0 Critical Infrastructure Improvement Issues in the Basic MAPP

5.1 Availability of funds for the poor

Funding for infrastructure improvement for the poor is assured for seven years, as long as the municipalities show adequate progress on reforms and prepare MAPPs. Judging from the present enthusiasm, these conditions will be fulfilled by most municipalities. The bulk of DFID infrastructure funding is designated for in-settlement improvements. The selection of priority settlements through the transparent process of analysis and decision making in the MAPP ensures that this funding will necessarily benefit the poorest communities in the analysis. Off-site infrastructure funds may lead to benefits accruing to the non-poor. However, guidelines specify that only up to 35 per cent of the infrastructure funds are to be used for off-site infrastructure, and that at least 50 per cent of the beneficiaries should be poor. This implies that, at the most, 17.5 per cent of the infrastructure funds may benefit those other than poor communities.

Subsequent cycles of MAPP are likely to include additional funds from other programmes. In any case funding for poverty programmes such as Swarn Jayanti Shahari Rojgar Yojana (SJSRY) and National Slum Development Programme (NSDP) will be available to the municipalities. Programmes and projects funded through the UIF will also complement MAPP funding.

The more ambitious and high-performance municipalities are expected to generate local resources, some of which may fund poverty-reduction initiatives. They may also use the Full MAPP to mobilize additional resources from national and international funding agencies, some of which have already shown a keen interest in the MAPP.

5.2 Targeting the poor for infrastructure improvement

Targeting of poor settlements in the MAPP is determined by two interrelated parameters: (i) poverty and vulnerability and (ii) infrastructure deficiency. Shortcomings, in quantitative and qualitative terms, in water supply, sanitation, roads, drains, flooding/erosion, street lights and garbage disposal are used as indicators. The Basic MAPP Guidelines give guidance on the selection of the indicators. It is left to the municipalities to give an appropriate weight, depending on the local conditions, to each of the indicators for arriving at a composite index of environmental infrastructure deficiency.

Poverty and vulnerability indicators are grouped into three broad categories: economic, social and entitlements. Initially a wide range of indicators were listed, based on earlier poverty data for the 32 towns,[7] participatory poverty studies elsewhere in the country and the findings of the first PPA. The list was finally shortened to two indicators: BPL (below poverty line) and SC/ST (scheduled caste/scheduled tribe) as the information on the other economic and social indicators was full of gaps. These two are also the conventional indicators used for targeting poverty programmes in India.

Settlements are given scores for each of the indicators. The average scores for poverty and infrastructure deficiency are then determined separately and clustered into three ranges representing the worst, not so bad and best settlements. These ranges of 1 to 3 are then used to rank poor settlements according to poverty and infrastructure deficiency and represent them on a simple matrix format. The settlements, that are the poorest and most deficient in infrastructure are in the cell in the bottom right corner (p3/i3). (See Figure. 2)

Figure 2: **Poverty and Infrastructure deficiency matrix**

Poverty and vulnerability (p) →

Infrastructure deficiency (i) ↓

p1/i1	p2/i1	p3/i1
p1/i2	p2/i2	p3/i2
p1/i3	p2/i3	p3/i3

Source: Based on Basic MAPP Guidelines (APUSP 2001).

The 3x3 matrix, as it is now called, has emerged as an appropriate tool for targeting poor settlements for infrastructure improvement in the MAPP. It gives a comparative picture of poor settlements in a town. It is simple enough to be understood and used as a working tool by a diverse group of stakeholders, including women representatives of poor communities. It is transparent and non-controversial and acceptable to local politicians, poor communities and municipal officials. The working group on poverty and infrastructure, which includes representatives from the major stakeholder groups, carries out the analysis. In all the MAPPs completed so far, the poorest and most deficient settlements were selected at the strategy formulation workshop.

The matrix can be considered as a major breakthrough, considering the conventional and accepted practices of either evenly spreading resources across all the slums, or selecting slums on the basis of slum notification and political constituencies. However, it has been criticized by academics for presenting an oversimplified picture of complex realities. Here it is argued that it is not the matrix but the data and indicators that need to be more robust. Using the existing data for the Basic MAPP has meant outdated information and a narrow range of social indicators. Fortunately, the municipalities and the MSU recognize this deficiency. Six of the eight municipalities have included accuracy of information on poor settlements in their reform proposals in the Basic MAPP.

Further, the PPAs are expected to demonstrate the importance of certain indicators and help resolve issues like whether or not entitlements are critical as indicators of poverty and vulnerability. The sharing of findings from the PPAs is an important step in understanding poverty in the Class 1 towns, and therefore, towards better targeting of the poor.

The other weakness related to targeting in the Basic MAPP also pertains to inadequate data. Data exists only for slums already notified under the AP Slum Improvement Act of 1956. Therefore, only these have been included in the Basic MAPP. Unnotified slums have been excluded, even though they are included in APUSP. Local knowledge confirms that usually the unnotified slums are much worse off, both in terms of poverty and infrastructure deficiency. However, it is the lack of data that prevents such settlements from being included in the first Basic MAPP.

The advantage can be taken here of MAPP as a rolling plan and of the opportunity to include critical inputs during the seven-year life of the project. Better understanding of poverty and up-to-date and more comprehensive data will certainly lead to better targeting of the poor in subsequent cycles of MAPP

5.3 The town-wide perspective on environmental improvement for the poor

The town-wide perspective is brought in through stakeholder workshops and the town-wide analysis of infrastructure and environmental problems. The technical feasibility checks of priority settlements are done with reference to service networks and site conditions. The requirement of producing an environmental statement for a town further strengthens the wider perspective. This kind of analysis is new to all the 32 municipalities; and here too, insufficient and inaccurate information has been a problem. But the process of analysis has created awareness in municipalities, and most of them have included related studies and information systems as part of their reform proposals.

One of the positive aspects of APUSP is the provision for linking slum infrastructure to town infrastructure. This way, linkages to water supply and drainage networks are assured.

However, higher levels improvements in networks, sources and disposal points are not included in the project, even though they may be necessary for the efficiency of services in slums. This requirement can only be addressed if municipalities are able to mobilize resources on their own strength or through funding institutions in later cycles of MAPP.

Even the identification of off-site infrastructure in the MAPP has not been without the problems of competing demands and the wish of politicians and taxpayers to take up general and more visible infrastructure improvement in the town. The design of off-site infrastructure in the first MAPP has proved to be a challenging task in the absence of suitable maps and other information with municipalities.

The other limitation is that though town-wide environmental hazards that have an impact on the poor are identified, there is no substantial way of addressing them through project funding. They are essentially seen as limitations to environmental improvement in poor settlements.

5.4 Participation of poor communities

Microplanning is now the accepted tool in India for eliciting the participation of poor communities in determining their needs and priorities and planning for the improvement of their own settlements. The difference here is that microplanning is nested within a larger town-wide framework of participatory analysis and decision making, unlike the usual slum-by-slum approach to participation, facilitated by municipal or NGO field staff.

The MAPP uses to advantage the federated structure of women's groups already existing in towns. The functionaries of the apex-level CDS participate in the MAPP formulation workshops side by side with the mayor/chairperson, councillors, the commissioner and officials of the municipality, MLA, MP and other institutions. They are also members of the working group on poverty and infrastructure deficiency. In addition, the CDS president is a member of the MRC. This way the MAPP process accords participation of poor communities a high status in decision making.

Experience so far has shown CDS members to measure up to the challenge of participating effectively with other stakeholders. This can be attributed to the fact that they represent the primary stakeholders and have been selected for their leadership qualities and the ability to negotiate and manage activities on behalf of the poor. They have also attended capacity-building workshops conducted as part of the SJSRY programme. The MAPP process itself is a capacity-building measure and gives all stakeholders an equal opportunity to participate.

Microplanning is the first step in articulating funding proposals for infrastructure improvement and is therefore given importance by municipalities. It has been designed as an interactive process between community groups and municipal field staff. Problem identification and demand assessment is led by RCVs and NHC convenors, who constitute the base and middle levels of the CDS structure, while the municipality responds with technical solutions for infrastructure. Finally a joint decision is made. This implies that both sides have to have the capacity to deliver their part of the results and work together.

Experience from the Basic MAPPs shows that representing their communities is a new experience for many of the RCVs. They are not even fully aware of their general roles and responsibilities. The brief training on microplanning is not enough of a basis for effective participation. This is expected to change once the capacity-building programmes of Component 3 reach the towns. However, that is unlikely to happen before the first cycle of Basic MAPPs is completed in the 32 towns.

5.4 Municipal capacity to support the process of infrastructure improvement and work with poor communities

The MAPP process has institutionalized multidisciplinary working in the municipality. However, the efforts need to be improved in subsequent cycles of the MAPP. Municipal micro planning teams, for instance, consist of engineering, sanitation and revenue staff, who are not used to working with communities. The brief orientation on microplanning is not sufficient to promote the skills and attitudes required to work with poor communities. Only 8 out of the 32 municipalities had community development staff till November 2001. The rest are in the process of being recruited. Project funds will support community development staff as and when they are put in place. It is assumed that they will be continued after the project period, as the higher revenues generated by financial reform will meet the expenditure towards their salaries. That, of course, may not happen in all towns.

Along with training and placement of community development staff, the capacity of municipalities to work with the poor is expected to be strengthened through appropriate communication and information tools and the expansion of the activities of the present citizens' service centres to include information services.

Another activity would be to develop the skills and attitudes of other staff, particularly engineers, for working with the poor. Microplanning, contract management, operation and maintenance of infrastructure -- all require engineers to work with community groups. This is seen as an important part of the municipal performance improvement activity. Yet other activity is the training of elected representatives to be more responsive to poor communities.

The inadequate capacity of the municipal staff to handle even simple technical work is a cause for concern. Not all the engineering positions in municipalities are filled because of the state government's ban on recruitment. The skills for analysis and design among the staff are also limited. The thoroughness in work practices brought in by APUSP is completely different from their previous experience. In the first cycle of the Basic MAPP, it is proving to be too demanding, in spite of training and consultancy support.

The importance of strengthening municipalities through various means cannot be overstressed. For this reason, training, appropriate institutional practices and adequate staff strength are seen as essential components of the reform process in APUSP.

5.6 Commitment to operation and maintenance of services

The municipalities have committed their own resources for the operation and maintenance (O&M) of services. These are expected to be strengthened once revenue improvement takes place as a result of reform. Making a provision for O&M in the municipal budget is a qualifying condition for the next round of

MAPP. Some of the O&M activities can be undertaken by the benefiting communities. This aspect has been introduced in the micro planning and needs to be taken forward during implementation. The institutional aspects of O&M also need further attention in the subsequent MAPPs.

6.0 Concluding Remarks: Responsive and Sustainable Environmental Improvement for the Poor in APUSP

The MAPP is a good basis for sowing the seeds of sustainable and responsive environmental improvement with the institutional focus of the municipality. The investment on infrastructure improvement forms the starting point for immediate municipal performance improvement in terms of working with the poor, improved analysis of problems and targeting, multidisciplinary working, and allocation for operation and maintenance of improved services. It remains to be seen whether or not the revenue increases as a result of reform are subsequently used for improving the conditions of the poor and for supporting institutional arrangements for working with the poor. A better-informed and vibrant civil society and transparent processes may actually allow that to happen.

The MAPP also institutionalizes and legitimizes the participation of women representatives of poor communities in the strategic and grassroots level decision-making processes, and establishes a link between the different levels. Transparent processes and the opening up of opportunities for the participation of a wide range of stakeholders are already beginning to have a spin-off in terms of creating a town-wide constituency for the poor. This will be taken further through civil society partnerships and networks. The functional and purposeful approach to participation in the MAPP needs to be supported by an awareness of the rights and responsibilities as well as the empowerment strategies of Component 3 for medium and long-term benefits.

The analytical and rolling nature of MAPP allows weaknesses in the areas of service delivery and working with the poor to either be addressed as part of the reforms component or to be taken up in the next MAPP cycle (e.g., data on poor settlements).

APUSP virtually places the municipality in the position of a 'mother institution' for poverty reduction. Not only is it expected to imbibe responsive procedures towards the delivery of services for the poor and their sustenance, but also be involved in the parallel civil society initiatives through the PPA and membership of the town level working group. This brings into focus the importance of municipal reform and performance improvement and links up directly with the policy perspective of the government on decentralization and municipal responsibility for poverty alleviation. It is for this reason that municipal capacity building is high on the agenda of APUSP.

The APUSP project principles, its multipronged approach and the down-to-earth mechanism of MAPP to translate these in the context of each municipality- all create a highly conducive environment for sustainable and responsive infrastructure improvement for the poor. The first year of project implementation has shown that APUSP certainly demonstrates a way forward.

7.0 Emerging Principles for Responsive and Sustainable Environmental Improvement for the Poor

Even though APUSP is yet to deliver the goods, it still provides some lessons for designing and undertaking responsive and sustainable environmental improvement for the poor. These can be articulated in a set of principles as follows:

- Committed funding for infrastructure improvement for the poor, linked to municipal performance
- Infrastructure improvement for the poor within a city-wide perspective of services and environmental problems
- Participatory and transparent decision making, involving poor communities
- Simple and clear procedures for participatory decision making
- Building municipal and community commitment to the operation and maintenance of services

- Capacity building of municipal staff for technical competence and working with the poor
- Filling the information gaps that hamper decision making

NOTES

1 A nation wide classification based on population. Class 1 towns have a population between 100,000 and 1 million.

2 The slum population is often used as a proxy for poor population, even though it is well established that not all people living in slums are poor.

3 A ward is the lowest administrative and electoral unit in a municipality.

4 Civil society organizations (CSOs), for the purpose of APUSP, include development NGOs, voluntary organizations, academic and research institutions, local media, community-based organizations, private sector companies, professional and business bodies.

5 The Government of AP has appointed a project coordinator and set up a municipal strengthening unit (MSU) and an appraisal and monitoring unit (AMU) under the Commissioner and Director of Municipal Administration to manage the first two components of APUSP. Component 3 (working with civil society) is managed separately by a management and support team (MAST). An empowered committee with state government and DFID representatives makes policy decisions.

6 Municipal corporations are headed by elected mayors and municipalities by elected chairpersons. Four of the 32 towns are corporations. A councilor/corporator in the municipal council/ corporation represents each municipal ward. The MLA (member of the legislative assembly) is the locally elected representative in the state government legislative assembly. The MP (member of parliament) is the local representative in the national parliament. The municipal commissioner is the executive head of a municipality and is appointed by the state government for a term of about three years.

7 Recent surveys of slums/ the poor in the 32 towns are: the AP slum survey (1994), BPL survey as part of the Swarna Jayanti Shahari Rojgar Yojna (SJSRY) programme (1998), and the multipurpose slum survey as part of the Janmabhoomi programme of the AP Government (2000).

REFERENCES

Amis Philip 2001, Rethinking UK Aid in Urban India: Reflections on an Impact Assessment Study of Slum Improvement Projects, *Environment and Urbanization,* Vol.13, No.1, April 2001.

APUSP 2001, *Basic MAPP Guidelines*, Government of Andhra Pradesh, Department of Municipal Administration and Urban Development, Hyderabad, available at http://www.apusp.org.

DFID 1999, *Andhra Pradesh Urban Services for the Poor (APUSP) Project Document*, unpublished.

DHV Consortium 1998, *APUSP: Draft Design of Institutional Reform and Urban Services Components,* unpublished report.

Ministry of Urban Affairs and Employment 1998a, *Guidelines for Swarna Jayanti Shahari Rojgar Yojna*, Government of India, New Delhi.

Ministry of Urban Affairs and Employment 1998b, *Guidelines for National Slum Development Scheme,* Government of India, New Delhi.

Ministry of Urban Development 1992, *The Constitution Seventy-Fourth Amendment Act 1992 on Municipalities*, Government of India, New Delhi.

Rao K Rajeswara 2000, "Clean and Green Cities: Participation of Communities/ Neighborhood Groups", *Shelter*, Vol.3, No.1, January 2000.

Rao K Rajeswara 2001, The Municipal Action Plan for Poverty Reduction (MAPP): A New Tool for Municipal Planning to Promote Participation, Transparency, Responsiveness and Strategic Vision , paper presented at the Launch Workshop for the UNCHS Good Urban Governance Campaign, September 2001, New Delhi.

13

Influence of Political Context on Municipal Efforts in Community Participation

Experience of Sri Lanka

Lalith Mallawarachchi

Background of the Participation Debate

Ideas on participation have been debated since the last few decades. The early discussions focused on the way communities interacted with the government in development projects. Arnstein[1] established the idea of a ladder of participation, which functioned as a continuum ranging from the most exploitative and disempowered to the most controlling and empowered. These ideas enabled theorists and analysts to describe various types of participation in terms of increasing degrees of decision making.

In 1983, Moser[2] proposed two distinctive means by which community participation takes place – first participation as a means; and second, participation as an end in itself. Participation–as-means implies that people are mobilized with the purpose of *achieving a desired outcome* (for instance, improved water supply or drainage). This is commonly evaluated in terms of the measurable outputs of the process. Participation–as-an end is not measurable in terms of development goals but in terms of devolution of power. It is a process where the *outcome itself is increasingly meaningful participation in the development process;* and where the real objective is to increase the control of marginalized groups over resources and regulative institutions. Moser (1983) suggested that the operational constraints on participation–as-a–means include inadequate delivery mechanisms, while the constraints on participation-as–an–end are structural and include institutional opposition to the redistribution of power and resources.[3]

This distinction has strongly influenced the catagorization of participatory projects. It was found to be helpful in efforts towards municipal capacity building for community participation and for discussion amongst municipal, community and NGO actors.

The conceptual framework for participation was described by Samuel Paul[4] in 1987. He identified five types of project objectives. *Cost sharing, efficiency, effectiveness, beneficiary capacity, and empowerment through information sharing, consultation, decision-making and initiating action.* His analysis identified three instruments of participation: *user groups, community workers/committees and field workers.*

Subsequently Schubeler[5] described that participation exists 'in a wide variety of forms, ranging from government involvement in community–based development activities to people's participation in government –directed management functions.'

More recently, Abbott[6] has stressed the potential of different types of participation and participatory processes in relation to the political context. He proposes that the duality of open and closed governmental environments fundamentally influences the entire concept of participation. Abbott proposes that participation in the context of democratic governments ranges from manipulation to community control.

Role of Municipality

Though there is a view that increasing government involvement will compromise the quality and output of participatory processes achieved by non-governmental actors through more focused and skilled initiatives, it is necessary to work with municipalities to bring about the scale of initiative likely to have a broader impact, in the arena of urban infrastructure. These elements develop more integrated solutions to service delivery and are crucial for effective participation and sustainable development. Moreover, pursuing community participation and community partnerships with municipalities is found to be a more effective approach to service delivery. Not only will top-down, supply-led approaches continue to produce inappropriate to unsustainable infrastructure, but the inertia of this perspective will continue to undermine the participatory efforts of other actors.

It is also necessary for the government to be aware of the risks of 'institutionalizing' participation. There is ample evidence of many government-led projects that never really absorbed the essence of participation-projects that were rushed through in order to spend budgets and those that replicated conventional methodologies. Further, many participatory processes have been pursued without attention being paid to the skills and organizational capacity of the actors responsible for the implementation process.

External Operation Context

Though the focus of the discussion is on the municipal level, which is responsible for the delivery of services and infrastructure within the scope of local government, it is crucial to identify the constraints on and support for participation created by external factors, which combine to form the municipal operating context. Therefore, in the current efforts of municipal decision-makers attempting to implement a participatory process, it is necessary to understand where the municipalities are placed in terms of participation, where they find support and resistance; what laws or policies influence their capacity; and what are the implications of the political environment for the emergence of participatory processes. The primary external factors likely to affect municipal capacity for participation include the political context, the legislative and policy framework, and the administrative context in which a municipality operates.

This paper discuss the influence of the political context, as an external operating factor, on the participatory process, with reference to the focus on the subject by Janell Plummer, based on a case study from Sri Lanka.[7]

Political Context

In many contexts, the key factor determining the formation of municipal-community partnerships is the political background of, and framework for, such action. Participatory processes that advocate a shift in existing power relations and a change in the structure of resource allocation, though not seen as harmless exercises in working with the poor, are considered as threats to the existing political environment and the status quo.

The political context influences municipal functions at a number of levels. First, the general political milieu will influence the scope of the participatory processes and the methodologies that can be used to implement them. A spectrum of political contexts exists: from one that is proactive and supportive to one that allows participation to proceed unabated but makes no attempt to facilitate the process. The former scenario is not common but is illustrated by case studies carried out in Colombo and Kerala, where politicians actively pursued participation in programmes, processes and institutions. The latter scenario is more common and frequently emerges in contexts where investment is conditional upon participation (such as donor-funded projects) and where municipalities are hoping to draw on community resources.

The second characteristic that affects participation is the dynamic nature of politics and government. Changes in political parties and ideologies can have a detrimental impact on projects driven by political motivation. While political support can lead to participatory endeavours, the perception that a project is politically aligned can threaten the sustainability of its approach if the leadership is removed. In Colombo, the setting-up of the community development committees (CDCs) and nurturing of the Community Action Planning (CAP) process over more than 15 years was rapidly undermined by a change in the ruling party at the national level. The incoming government withdrew support for the participatory methodology and insisted on the disbandment of CDCs.

There is also a general conflict of interest between CDCs and politicians. In most situations, the politicians of whichever party in power wish to be seen as the main benefactor - the "boss" who provides people with land, housing, water, clinics, sports facilities, etc. In return, he or she wants public acclaim and votes. CDCs are seen as a threat to these vertical relations of political patronage because they too can take credit for community improvements and can provide a focus for cooperation between people of different backgrounds and political affiliations. 'Horizontal' linkages of this nature can increase community capacity to bring out service improvements without reliance on political gifts. CDCs, like other CBOs, are perceived as a threat to vertical patron-client relations.[8]

In the context of a tiered system, potential conflicts between levels of government can also undermine participatory processes. Deteriorating relations between the state and the local levels of elected government heighten blockages and political manoeuvring. In the initial stages of the Colombo Urban Housing Subprogramme, each governmental tier was controlled by the same political party. In the latter stages, conflict between the municipal-level government and the national-level government constrained effective action in all aspects of poverty reduction.

The Impacts of Political Change

In Sri Lanka, an enabling approach with increased community participation was significant in the policies and programmes for environmental improvements in poor areas during the period 1979-94. The prevailing political context encouraged the adoption of such an approach. It was characterized by the CDC structure and the CAP methodologies, which were promoted at the national, provincial and local levels of government. The success of the approach was highlighted by the supporting government action and the implementation through the Million Houses Programme, which established it as a forerunner of numerous participatory pilot projects in South and South-East Asia.

The experience of community participation in Colombo or elsewhere in the country since then is a reflection of the impact of the political environment. Up until 1994, the same political party was in control of different tiers of the government, i.e., national, provincial and local (provincial councils were formed in 1987 with the 17th Amendment to the Constitution). This ensured unified efforts towards implementing policies and programmes and particularly enhanced the effectiveness of the participatory process. However, the change in the ruling party at the national level during the general election of 1994 saw a reexamination of political ideas and priorities. Immediate conflict with the political objectives of the CDC structure was observed, resulting in the withdrawal of political support for the committees, the community action plan approach, and the institutional framework supporting the process. This was found to be the primary factor in the gradual collapse of the CDC movement. The emergence of what was recognized as a sustainable and successful participatory approach over a decade was completely undermined by the said withdrawal. In the new political environment, the Colombo Municipal Corporation (CMC), controlled by a different political party, faced numerous constraints, leading to shrinkage of municipal activity with communities.

The impact of the eroded national-level support after the 1994 election was exacerbated by the strained relations between the provincial and municipal levels of government. Problems in the relations between the Western Provincial Council (WPC) and CMC have been developing since 1996. Delaying the approval for CMC initiatives has been common with the WPC, particularly when the delays were likely to prevent the CMC from deriving political mileage.

The handing over of responsibility for supervising the management of local authorities to the provincial councils created more problems for CMC and further undermined the participatory process. WPC, as well as the other councils formed, failed to fulfil the legal obligations of decentralization to share power and strengthen the capacity and resource base of municipalities, including the CMC, within its jurisdiction. It has yet to enact important statutes to provide the local-level CMC with operational authority. As a result, the Municipal Corporation Ordinance is the applicable law, which places limitations on the functions of municipalities.

Some of the key activities managed by the provincial government are local government, health and education. The WPC however, lacks the technical capacity, policy coordination and resources necessary for the implementation process. Central government agencies, which have at their disposal greater information and expertise, often bypass the WPC in these activities.

CMC has a key role in improving the city environment in terms of municipal administration, but it does not have the necessary control over the main service provision activities in Colombo. Decision-making authority for four important services (water, sewerage, electricity, housing for the poor and drainage) is vested at the national level with central government agencies, which are mostly controlled by the Ministry of Housing and Urban Development. Furthermore, the activities of these agencies cannot be influenced by CMC or even WPC. They are frequently seen as encroaching upon the activities of CMC.

Another important aspect in this context is the political manoeuvring of some kind or another, which have been witnessed at the microlevel. Success in these approaches during 1979–1994 were often the result of the active efforts of local politicians to lead community participation and the CDC movement. Shifting political objectives since the mid-nineties destabilized local-level support. "The change from ward committees to promotional representation also reduced the accountability of elected councilors to the electorate, thereby reducing the lobbying of the urban poor."[9]

Municipal action towards participation does not operate in a void but in a broader context determined by external factors. Political, legislative, policy and administrative factors have a significant impact on the capacity of municipalities to meet their objectives. Hence, the objectives and processes need to be considered in relation to the constraints of and support from the external operating context, which is largely outside the sphere of influence of the municipality. It is critical to identify the key organizations and individual actors who lead that policy context and their potential impact on municipal-levels efforts.

NOTES AND REFERENCES

1 Arnstein S R 1969, "A Ladder of Citizen Participation", *Journal of American Institute of Planners,* Vol., no. xxxv, p.19).

2 Moser Caroline 1983, *Evaluating Community Participation in Urban Development Projects,* Development Planning Unit of University College London, Working Paper. No 14

3 Plummer Janelle 1999, "Municipalities and Community Participation", *Sourcebook for Capacity Building*, Earthscan. London.

4 Paul S 1987, "Community Participation in Development Projects", World Bank Discussion Paper. No. 6.

5 Schubeler, "Participation and Partnership in Urban Infrastructure Management", UMD Policy Paper No. 19, World Bank, Washington.

6 Abbot J, *Sharing the City*, Earthscan, London.

7 Plummer Janelle 1999, *Municipalities and Community Participation, Sourcebook for Capacity Building*, Earthscan. London

8 Fernando Austin, Steven Russel, Anoushka Wilson and Elizabeth Vilder, *Urban Governance: Partnership and Poverty, Colombo,* University of Birmingham, Working paper 9

9 Plummer Janelle, 1999, *Municipalities and Community Participation, Sourcebook for Capacity Building*, Earthscan. London

14

Strategic Planning Approach for Urban Environmental Management at the Local Government Level

Experiences of Sri Lanka and Thailand

L A S Ranjith Perera

1.0 Introduction

Local government institutions in developing countries are confronted with ever-growing urban problems on the one hand and the demand for quick and reliable solutions to those problems on the other. Within the limitations of available resources and capacities, it is nearly impossible to find solutions to all these problems, which are increasing in number and magnitude. Therefore, utilization of the limited resources of local governments has to be backed by rational decision making, which takes into account the priority needs of its citizens.

Strategic Planning is a technique to prioritize problems and arrive at the best possible solutions within the framework of limited resources and capacities. Moreover, it is a systematic way to manage change while seizing the opportunities emerging through the very process of change. It is also a creative process to identify the priority needs of citizens and stakeholders and fulfil most of them through clever strategies and actions. In other words, strategic planning focuses on the allocation of limited resources to critical areas (PTI 1983).

Strategic Planning (SP) has been used as a technique to manage change in large corporations and businesses since the 1960s. In order to face the rapid changes of macroeconomies both at local and international levels, these establishments from time to time reassess their internal strengths and weaknesses and foresee external opportunities and threats (SWOT analysis), before implementing new strategies to reposition themselves in the competitive business environments. The success of SP first in private sector organizations, and later even in non-profit organizations such as national universities, has prompted urban authorities to borrow and adapt this technique for a new paradigm that views city as a large enterprise, which is primarily a place for earning and capital formation. In this paradigm of the 1990s, cities attempt to reposition themselves to face the challenges posed by other cities, which are competing for opportunities, and also reorient themselves to meet the effects of globalization. Presently, SP has become a technique for improving the competitive position of a city or municipality, much as it is used to improve the competitive position of a corporation, by evaluating how it stands in relation to changing markets and competition. By now, most urban authorities have realized that conventional urban planning techniques, such as master planning and structure planning, are no longer effective in the context of high competition and constantly changing market conditions. The implementation of these techniques requires long time frames, of 20-30 years, colossal amount of resources, and consistent development policies by governments. Moreover, most grand plans have failed to achieve their objectives because they have been conceived against the wishes of the stakeholders by unrealistic and ambitious planners.

In contrast, by viewing cities as corporations that need to be run like businesses or managed like companies SP has resulted in a major shift in the planning paradigm. This is a major deviation from the "functional efficiency of cities" emphasized in the master planning paradigm or from the "integrated development of cities" emphasized in the structure planning paradigm. The past experiences of master planning and structure planning have revealed that such plans do not have the flexibility to repeatedly adjust and readjust to the rapid changes in the plan area that occur even before the actual implementation of such plans. In other words, long-term plans become easily redundant in the context of rapid urbanization and complex urban problems. This has lead urban planners as well as local authorities to search for a development approach, that can address the critical issues and identify effective and acceptable solutions within limited resources and a shorter period of time. Although participatory form of urban planning came to the fore as an alternative approach after its initial success in the rural sector, the application of this technique in urban areas became substantially more problematic because of weak social cohesiveness and the continued influence of local government institutions (Mitlin & Thompson 1995). If participatory methods of urban planning are to work, fundamental changes will be required in the management practices of local government agencies (Atkinson & Vorratnchaiphan 1996:236). The urban management paradigm emerged in the early 1990s as a more effective approach to urban development in this context.

In parallel, the focus of donor agencies also began to shift from planning-oriented solutions to management-oriented solutions as they realized the difficulty in planning for mega urban areas to solve dynamic and multifaceted sets of problems. Presently, several development programmes operating in the Asian region and funded by multilateral agencies, viz., the Urban Management Programme (UMP-Asia supported by UNCHS/UNDP/ World Bank and the City Development Strategies (CDS) Programme of UNCHS/UNDP/World Bank, use SP as more of a management technique to identify development priorities and implement action plans/projects. A notable shift in both these programmes is direct intervention at the local government level without channeling development assistance through national-level ministries or agencies. This is seen as an attempt by donor agencies to help in capacity building for urban management at the local government level.

In countries with mega cities, national-level agencies and city governments (e.g., the National Economic and Social Development Board (NESDB), and Bangkok Metropolitan Administration (BMA) of Thailand) have also independently effected a shift from structure planning to Strategic Planning in recent times due to the sheer complexity of the urban problems they are faced with (NESDB-DANCED-Kruger 1996 & BMA-MIT 1996). SP has helped both these organizations to identify critical problems, which need solutions on a priority basis. Megaurbanization has made comprehensive planning of mega cities a near impossible task. As an alternative, more strategic solutions to deal with the problems of mega urban regions are coming to the fore. At the other end of the spectrum, small local authorities with limited resources and capacities also find SP to be an effective technique, which can deliver more satisfactory results to the citizens. The UMP-Asia programme noted above covers urban areas right across the spectrum, including Mumbai of India with 15 million people and Phuket Island of Thailand with 70,000 people.

The intention of this article is to highlight the suitability of the strategic planning technique for achieving even more specific goals such as urban environmental management at the local government level. The paper draws on some of the experiences of Sri Lanka and Thailand in the course of their participation in the above-mentioned programmes in order to illustrate the effectiveness of the technique for local government bodies such as municipalities.

2.0 Strategic Planning for Urban Environmental Management

As in the case of general urban problems, the increasing scale and complexity of urban environmental problems in many developing countries have reached a point where there is a manifest need for a more strategic approach to identify the priority areas for intervention. Urban environmental problems need to be comprehended holistically in order to identify the critical issues, instead of offering end-of-pipe solutions in an ad hoc manner. The strategic planning methodology requires a complete scanning of the environment before identifying the priority issues (see Fig. 1).

The strategic approach towards urban environmental management that is derived from the generalized strategic planning technique has the following characteristics:

1. It focuses on prioritized environmental issues/problems.
2. It works within the framework of resources available/obtainable to solve environmental problems.
3. It recognizes multiple stakeholders in environmental systems, resources and commons.

Figure 1: Generalized Strategic Planning Process

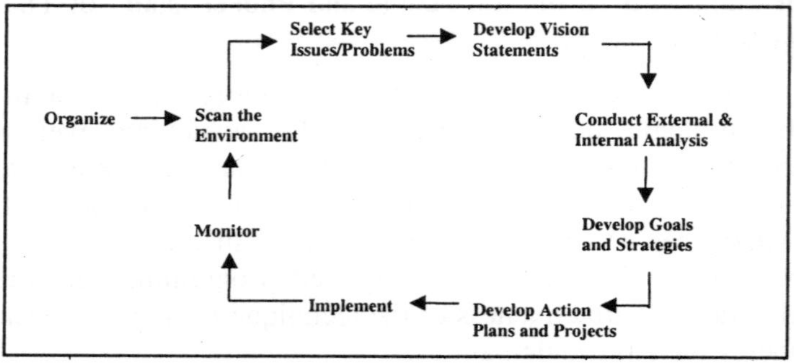

Source: PTI (1983)

4. It seeks to mobilize social initiatives on environmental protection through guided actions.
5. It develops a shared vision as a goal to achieve, with the consensus of all stakeholders.
6. It assesses the strengths and weaknesses of the context of intervention to achieve the goal.
7. It evaluates the opportunities available to be seized for achieving the goal.
8. It also evaluates the external threats facing those opportunities.
9. It proposes both short- and long-term strategies to achieve the goals, with strong emphasis on actions and results.
10. It seeks environmental sustainability through capacity building at the local level.

Unlike master planning and structure planning, SP is largely a bottom-up approach. Accordingly, structure planning in urban environmental management also focuses on the environmental issues prioritized by stakeholders rather than on specific problems of environmental media (i.e., air, water and land) as commonly highlighted by technocrats. Structure planning recognizes urban environmental management as a people-oriented process, not as management of environmental media and living organisms. Hence, the mode of operation of structure planning is essentially consultation with the stakeholders. It requires both public and private stakeholder involvement in the process. Private stakeholders here refer not only to the private sector (usually industry, banking, trade and commerce) but also to citizens, community organizations, non-profit institutions and others outside the public sector. Thus, strategic planning is a powerful technique for developing strong public-private partnerships at the local government level. It can produce better coordination and concerted actions within a local government area by enabling its citizens and other stakeholders to act with a view to achieving the same environmental goals. A strategy development effort is an opportunity to create and share a mutual vision of stakeholders about city development within a defined time frame. In formulating a shared vision, the stakeholders are asked: *Where are we now*, and *Where do we want to be in 5-10 years' time*. The answers to the second question are synthesized into a shared vision, which becomes a consensus goal for pursuit by individuals. A strategic approach to urban environmental planning and management calls for the attention of two major clusters of tasks; (i) establishing an agenda of the environmental issues/problems that need to be prioritized and the aspects or sources that need to be tackled in each urban setting; and (ii) selecting key initiatives and actions (see Fig. 2). These two broader tasks are handled in two separate phases in the strategic planning process. The specific tasks that each phase covers are as follows.

Phase I: Environment Agenda

1. Priority Environmental Issues

(a) Scan the Planning Area: Assess environmental issues and their impact at the local level, assess their relative magnitude, identify key factors and trends important for the future, determine how external forces influence the planning area.

(b) Select Key Issues/Problems: On the basis of the scan, build consensus and choose a few issues/problems whose successful resolution is critical.

Figure 2: Components of the Strategic Framework and Process

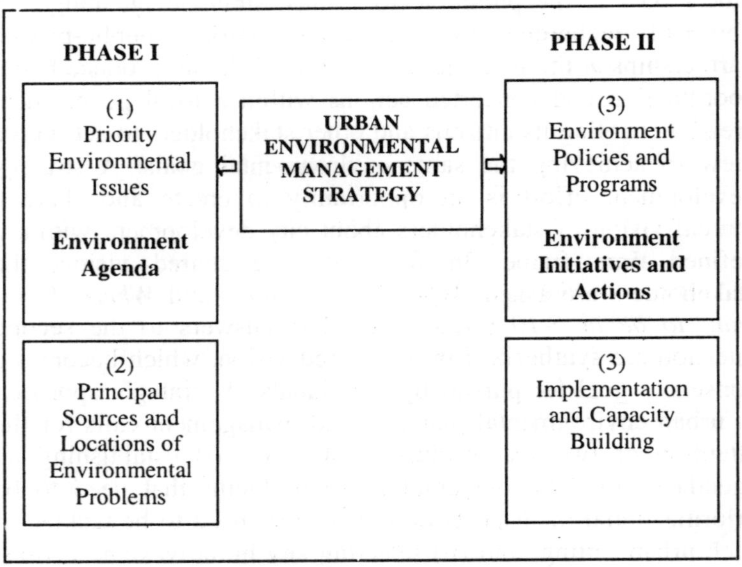

Source: Adapted from NESDB-DANCED-Kruger (1996:2-3)

(c) Set Priorities: Build consensus among the local stakeholders to select priority issues/problems.

II. Principal Sources and Locations of Environmental Issues/Problems

(a) Identification of Major Sources: Identify and analyze the sources of the problem within chosen critical issues.

(b) Consensus Building on Sources and Locations to be Targeted for Action

Work towards agreement of the stakeholders on major problems, the culprits and their location.

Phase II: Environment Initiatives and Actions

11. Environment Goals, Policies and Programmes
(a) External and Internal Analyses
Take an in-depth look at the external forces affecting the achievement of goals and policies; identify strengths and weaknesses of the planning area along with the availability of resources.

(b) Develop Goals, Objectives and Strategies
Based on the external and internal analyses, decide what can be achieved with respect to each issue and how it will be achieved.

12. Implementation and Capacity Building
(a) Develop Implementation Plan
Be specific about the timetables, resources and responsibilities for carrying out strategic actions.

(b) Joint Actions with Private/Civic Organizations
Share responsibilities with stakeholders in implementing action plans.

(c) Build Local Management Capacities
Build the capacity of both local authorities and stakeholders to sustain the strategic plan.

(d) Monitor, Update and Scan
Ensure that strategies are implemented, adjust them if necessary, be prepared to update when a major change occurs in the environment, be transparent in monitoring and evaluation.

(Adapted from PTI 1983 & NESDB-DANCED-Kruger 1996:2-3)

The generic form of structure planning has been further simplified for urban environmental management purposes in the local environmental action planning (LEAP) technique. This technique is recommended for local areas such as municipalities seeking to address priority problems of sustainable development

(Leitmann 1999:143). The technique involves several activities, which in fact are simplified steps of the strategic planning process (compare Figure.1, Figure.2 and Figure.3). In other words, the nine steps of the generalized strategic planning process are compressed in to three steps in the LEAP process. They are discussed below.

1. *Informed Consultation*

   Involve key stakeholders and obtain commitment from the political leadership, conduct rapid environmental assessments, clarify major environmental issues, set priorities for further action through an informed consultative process.

2. *Formulation of Local Environmental Action Plan (LEAP)*

   Set goals, phase out targets, identify strategies and solutions that cut across the concerns of the majority of stakeholders, prepare action plans based on least-cost options, make necessary institutional improvements for plan implementation.

3. *Implementation of LEAP*

   Initiate agreed programmes and projects, reform policies if necessary, solidify institutional arrangements, monitor and evaluate the implementation process, use the feedback on the results to adjust strategies and solutions.

Figure 3: Local Governmental Action Planning Process

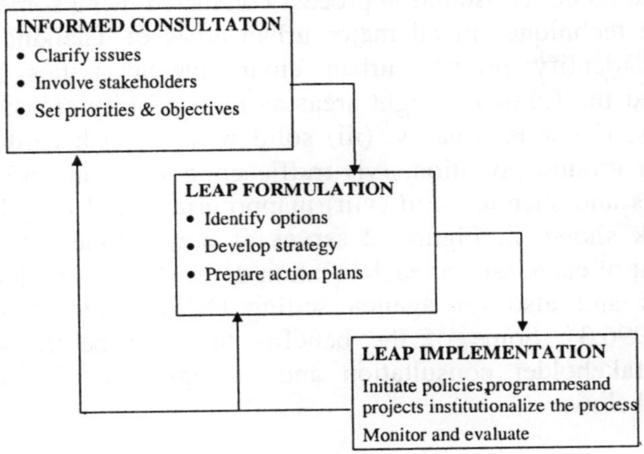

Source: Adapted fromLeitmann(1999:143).

As is evident from the preceding discussion, the success of the strategic approach to urban environmental management largely rests on the inclusion of multiple stakeholders in environmental assessment, decision making as well as strategy implementation. This requires a shift from planning and decision making at the top management level towards a more open process of public participation, stakeholder consultation and mutual decision making. This can happen in several ways:

1. conducting individual interviews with all stakeholder groups
2. conducting collective discussions with the representatives of all stakeholder groups
3. holding an open forum for local citizens, government organizations, private sector organizations, NGOs, CBOs and other interest groups
4. organizing key groups for smaller more spontaneous, discussions (e.g., at the neighborhood level)
5. conducting opinion surveys at the household or community level

6. using participatory environmental assessment techniques and coded scales (e.g., radar and bar graphs)

A stakeholder consultation process conducted using some of the above techniques in all major urban areas of Thailand in order to identify priority urban environmental issues has highlighted the following eight areas as high priority areas: (i) air quality, (ii) water quality, (iii) solid waste, (iv) hazardous waste, (v) ground pollution, (vi) traffic congestion and noise, (vii) slums and shanties, and (viii) inappropriate land use. The framework shown in Figure. 2 serves as a guideline for the assessment of each issue at each location depending on the local conditions and also for agenda setting (NESDB-DANCED-Kruger 1996:3). Some of the benefits that can be derived through stakeholder consultation and strategic planning are discussed below.

1. Accomplishing the important things

Immediate problems and demands often stand in the way of action on more important issues. As in other organizations, municipalities also tend to delay actions on long-standing problems and resort to ad hoc solutions often due to resource constraints. Strategic planning is a tool for generating action on high-priority items. A strategic plan answers the question "What action will make a difference in five or ten years from now?"

2. Community education and consensus building

The strategic planning process identifies trends and directions that shape the community. Recognition of these trends is an important first step toward agreement on the best response. Improving the public's understanding of the local situation can have other benefits as well.

3. Developing a shared vision

Strategic planning helps to define a new identity and establish a sense of direction. A sense of identity is important to glue together diverse elements of the urban community. (For e.g., the shared vision of Colombo: "A model city in Asia where people enjoy quality living, acquire optimal capacity and opportunity for creation of wealth for the well-being of all, sustain and improve the environment", (Fonseka-Sevanatha URC 2000:23).")

4. Positioning the city or municipality to seize opportunities
Strategic planning helps a city or municipality to become more proactive and to create its own future. It helps a community to identify its advantages in relation to the external environment, its region, state, nation and world.

5. Shedding new light on important issues through rigorous analysis
Strategic planning is a structured, disciplined process that can generate new information or draw new attention to important information that has been overlooked. It provides an integrated picture of the community's current position and future prospects. Strategic planning is also a useful technique for broadening one's understanding of the available resources. It offers leaders a chance to look beyond the day-to-day concerns.

6. Helping to identify the most effective use of resources, including public funds
Strategic planning supports targeting of resources to the most important problem and opportunities. It focuses on a limited number of key issues and establishes priorities. It emphasizes explicit consideration of resources and environmental forces. It avoids the frequent planning error of determining at the top decision-making level what ought to be done. Instead, it concentrates on the doable in realistic terms. This emphasis on resource availability means that strategic planning can be a means of determining how to use a limited supply of resources. In fact resource limitations appear to be one reason for the recent interest in strategic planning among the local leadership faced with an economy in transition.

7. Providing a mechanism for public-private corporation
Strategic planning for communities requires inputs from key decision makers and opinion leaders representing major

segments of the community. Bringing these people together in a constructive effort may be extremely valuable.

However, in reality the strategic approach for urban environmental management has produced mixed results in developing countries.

3.0 Some Experiences of Utilizing the Strategic Planning Technique for UEM

Strategic planning as a mode of operation is used in the two Asian regional projects noted above, viz., the UMP-Asia project and the CDS Programme --- both funded by UNCHS (Habitat), UNDP and World Bank. Both the programmes are very similar in their approaches and operational areas. The former, which operates in 20 cities in 11 countries focuses on urban poverty alleviation, urban environmental management, and participatory urban governance at the local level. The latter, which operates in three Asian cities, focuses on environmental planning and infrastructure development, urban poverty and social development, economic development through public-private partnership, and reforms in administration for good governance, also at the local level (UMP-Asia 2001:1, Perera and Sevanatha 2000:24). Both these ongoing programmes have given due emphasis to environmental planning and management and both follow a bottom-up approach, driven by a stakeholder consultation process operating at the local government level. The underlying premise of UMP city consultation is that poor city administration and environmental problems are often the result of weak rapport with stakeholders, particularly where bureaucratic and unresponsive modes of administration are the norm. The UMP city consultation approach has been designed to bridge this gap so that the city administration and key stakeholders in civil society can participate in decision making. The aim of a city consultation is to institutionalize participatory decision making in strategic planning and development. City consultations bring together local authorities, the private sector, community representatives and other stakeholders within a city or municipality to discuss specific issues and possible solutions to key urban problems.

Among the five cities most often mentioned in the success stories of the UMP-Asia project are Colombo municipality of Sri Lanka and Phuket and Nonthaburi of Thailand (UMP-Asia 2001). The success of the UMP-Asia project in Colombo led to its selection (together with its adjacent municipalities of Kotte-Sri Jayawardenapura and Dehiwela-Mt.Lavinia) as one of the three pioneering cities for developing a City Development Strategy under the UNCHS-World Bank Cities Alliance Initiative in 1999. Similarly, the success of the Phuket and Nonthaburi projects are replicated in the Yala Municipality in Southern Thailand.

3.1 Experience of Sri Lanka

The highlight of the UMP-Asia experience in Colombo is participatory urban governance. This was possible perhaps due to a change in the leadership of the municipality in 1997, which provided an opportunity for the Colombo Municipal Council (CMC) to implement innovative approaches towards good governance. Even before the launch of the UMP-Asia project in Colombo, the mayor had formed 10 teams of honorary advisors comprising professionals, entrepreneurs and leaders of civic groups to advise him on major issues of the city. Such initiatives of the mayor on participatory governance made for the easy implementation of the stakeholder consultation and strategic development process by UMP-Asia in 1998.

The series of stakeholder consultations organized by Sevanatha Urban Resource Center, the local partner institute (LPI) of UMP-Asia-Colombo, highlighted an urgent need for the revival and strengthening of community development councils (CDCs) and other community-based organizations, which were established in the 1980s and became defunct later. The LPI played a catalytic role in stimulating a more participatory form of urban governance involving the citizens. It organized a broad-based consultative process to engage a wide cross section of stakeholders in a "dialogue of strategic development". The series of city consultations held subsequently lead to the prioritizing of four major issues in the city: (i) enhancing the capacity of the CMC for revenue collection and mobilization, (ii) environmental management, (iii) increased participation of citizens in decision making, and (iv) responsive urban governance.

The achievements of the stakeholder consultations and the subsequent strategic planning and development process were:

1. Widespread consensus on the need for civic engagement in the basic services delivered and other development programmes
2. Revitalizing CDCs for environmental management and community contracting for the delivery of basic services
3. Increased involvement of elected councillors in development programmes and environmental management
4. Steps initiated for strengthening the district offices of CMC, which are responsible for the delivery of environmental services at the level of municipal wards
5. Donor consensus and increased coordination, especially for environment and poverty-related interventions in Colombo
(UMP-Asia 2001)

Success on these fronts was to do with the fact that the political leadership of the municipality already was inclined towards participatory governance and the organization already possessed a strong team of committed officials and a reasonable network of social and physical infrastructure. In other words, in the absence of political willingness and leadership and the strong capacity of the local authority, such achievements would have been not possible.

This was evident when the CDS Programme was implemented in 2000 in the adjacent municipalities of Kotte-Sri Jayawardenapura and Dehiwela-Mt.Lavinia. The former was promoted to the level of municipal council only in 1999, by which time it was plagued by inefficiency and the lack of human and physical resources. There were frequent conflicts between the local government and the citizens over environmental issues such as garbage disposal, poor drainage, illegal land reclamation, frequent flooding, mosquito menace, etc. Being the seat of the national government and the administrative capital of the country, it was a hot bed for political rivalries. Political parties have often tried to capitalize on environmental issues for their own gains. These rivalries were visible even during the consultation process, which eventually affected consensus building among the stakeholders. Since consensus among the stakeholders is fundamental to the success of strategic planning for urban environmental management, the implementation of CDS in the Kotte-Sri Jayawardenapura municipality was not as effective as in the Colombo municipality.

The experience at the Dehiwela-Mt.Lavina municipality was even worse. Although was been established 40 years ago, it has a very low capacity and efficiency compared to the Colombo Municipal Council. The municipal officials as well as most elected councilors shared little enthusiasm for the participatory form of governance and had no motivation to take on responsibilities in addition to their routine functions. Except on a few occasions, participation of the stakeholders in the consultation meetings was poor. Lack of stakeholder participation was more acute in this municipality. Although the political leadership showed some interest, it was not as strong as in the Colombo Municipal Council.

These experiences pose several questions. Is the strategic development process feasible in local governments where capacity is considerably low and the attitude of the political leadership towards participatory governance is lukewarm? Is prior capacity building necessary to implement a strategic plan at the local authority level since the methodology emphasizes the need for short-term actions. The overwhelming success of the strategic approach in Colombo and the unimpressive results of its adjacent municipalities can be attributed only to the relative capacities of the municipalities in terms of political leadership, revenue base, human resources and urban structures.

3.2 Experience of Thailand

The Bangkok Metropolitan, which is already a city of 10 million people (living in 1,565 sq km), is one of the mega cities included in the UMP-Asia Project. However, the planning process implemented at the Bangkok Metropolitan Administration (BMA) has differed from the UMP's general approach. It has not been either a participatory process at the organizational level or an interactive civic consultation at the community level. On the contrary, it has focused on the capacity building of BMA and its high-level officials. The wealth and power of BMA and the higher status of Bangkok in Thailand's national urban context may indicate its readiness for more integrated or comprehensive planning, instead of venturing into participatory planning. This is evident from BMA's attempt to prepare a comprehensive development plan in consultation with a team of professionals from USA, outside the scope of the UMP-Asia project. However, BMA soon realized the inappropriateness of the development vision and the strategic development plan prepared by the team of consultants. The plan, which is now shelved, focused more on addressing the physical development issues of the city rather than strategic issues, particularly environmental management issues, although the environmental problems of Bangkok are more chronic. Even the present strategic plan of Bangkok, which has been formulated by the planning professionals of BMA, has not deviated much from the physical infrastructure development-oriented strategies. Both are top-down strategic plans prepared without any stakeholder consultations.

In contrast, UMP-Asia's approach in Phuket and Nonthaburi was directly focused on the urban environmental management issues. As one of the premier tourist destinations in Thailand, Phuket has been at risk of suffering from the unplanned and unregulated development that has characterized growth in other top tourist destinations in the country. Concerned about the growing threat to the natural environment caused by increasing tourism activities and agro-industries in and around the city, the Mayor of Phuket, together with UMP-Asia and local partnering institutions have initiated a city consultation process to identify strategic directions that the island resort should take to sustain its economy. As a result of the consultation process, the local authority of Phuket showed keen interest in restructuring the municipal environmental management committee to include more local community and private business representatives. The major objectives of the local authority were to develop an urban environmental policy consistent with provincial and national development plans; undertake an urban environmental planning and management process; develop a continuous human resource capacity building programme; and strengthen community organizations. Among the outcomes of the stakeholder consultation and strategic planning in Phuket were:

1. An environmental vision and a policy has been formulated covering five priority areas: (i) built environment, natural resources, and land use; (ii) traffic, air and noise pollution; (iii) water supply, drainage and water pollution; (iv) community environment and environmental health; (v) solid and hazardous waste.
2. A strategic plan has been formulated to translate the policies into effective measures, including the drafting of an action plan, promotion of participation by diverse groups of stakeholders, dissemination of knowledge and information, establishment of networks, reforms in municipal management systems, and improvement in personal capabilities.
3. A capacity-building programme for municipal officials has been initiated, based on a self-conducted training needs assessment and covering topics such as planning, environmental management, sustainable development, and cooperation among agencies in environmental management.
4. Waste recycling pilot programmes have been initiated, combined with environmental education through women's groups.
5. Related community improvement projects have been initiated, such as the adoption of a self-regulating system for food vendors, hoteliers and small entrepreneurs.

The Phuket experience of urban environmental management through the strategic planning approach was first replicated in Nonthaburi, which is a part of Bangkok Metropolis, and then in Yala, which has already acquired the reputation of being one of the well-planned cities of Thailand. While the UMP experience in Nonthaburi was a success similar to Phuket -- mainly due to the enthusiastic participation of stakeholders, their commitment, and the sense of ownership that they feel about their living environment -- the experience in Yala has so far been hampered by a weak administration, the lack of enthusiasm and capacity of the local government officials, and difficulty in mobilizing the stakeholders. In particular, local government staff has shown little enthusiasm to take on what they view as responsibilities in addition to their routine functions.

These two later experiences of Thailand also confirms the earlier findings of Colombo that the strategic planning approach at the local government level is successful when the local authority already has some capacity and willingness to engage in stakeholder consultations.

4.0 Conclusion

In both countries, the success stories are from more established, medium-sized municipalities, not from larger or smaller local authorities. It was found that the more established and powerful metropolitan authorities, like BMA, had ignored the stakeholder consultation in preparing strategic plans, while the weaker and smaller local authorities, like the Dehiwela-Mt.Lavinia and Yala municipalities, have found it difficult to even hold consultations with at least key stakeholders. Therefore, the strategic plans that they had prepared lacked credibility in terms of catering to the needs of the real stakeholders of the city. Thus, strategic planning as a technique for urban environmental management appears to be suitable for medium-sized local authorities. In mega cities like Bangkok, urban environmental management has been given a low priority compared to urban management and infrastructure development; while in the case of smaller local authorities, capacity building needs take precedence over urban environmental management issues. Thus, it is apparent that for strategic planning to become an effective technique for the purposes of urban environmental management, reasonable capacities need to be developed by the respective local government institutions. General options for capacity building require to focus on: financial resources, human resources, training, exposure, technical assistance and equipment. However, the most powerful form of institutional capacity building for urban environmental management is the development of partnerships between key stakeholders. The lessons from Sri Lanka and Thailand amply highlight the crucial importance of this vital ingredient.

REFERENCES

Atkinson A. and C P Vorratnchaiphan 1996, A Systematic Approach to Urban Environmental Planning and Management: Project Report from Thailand, *Environment and Urbanization*, vol.8, no.1, pp. 235-48.

BMA-MIT 1996, *The Bangkok Plan – A Vision for the Bangkok Metropolitan Administration Area*, Bangkok Metropolitan Administration and Massachusetts Institute of Technology, Bangkok.

Fonseka, L. and Sevanatha URC, 2000, *City Development Strategy: Colombo*, Colombo Municipal Council, Colombo.

Leitmann J 1999, *Sustaining Cities: Environmental Planing and Management in Urban Design*, McGraw Hill, New York.

Mitlin D and J Thompson 1995, Participatory Approaches in Urban Areas: Strengthening Civil Society or Reinforcing the Status Quo?, *Environment and Urbanization*, vol.7, no.1, pp. 231-50.

NESDB-DANCED-Kruger 1996, *Urban Environmental Management in Thailand: A Strategic Planning Process – Final Report.* National Economic and Social Development Board-Danish Cooperation for Environment and Development-Kruger Consult, Bangkok.

Perera, L A S R and Sevanatha URC 2000, *Proceedings Report of the Consultation Process and Outline Development Strategy for Dehiwela-Mt.Lavinia Municipal Council Area*, Sevanatha Urban Resource Center, Colombo.

PTI 1983, *Strategies for Cities and Counties: A Strategic Planning Guide*, Public Technology Inc., Washington DC.

Sattam A. et al. 2001, *Final Report of Yala City Consultation,* Sukhothai Thammathirat University.

UMP-Asia 2001, *UMP*-Asia *Consultations*, Urban Management Program Regional Office for Asia and the Pacific, Asian Institute of Technology, Bangkok.

15

Urban Governance and Role of the Community in Solid Waste Management for Dhaka City

Golam Rahman

Introduction

Solid waste management (SWM) is today considered to be one of the most severe environmental problems of urban governance in Bangladesh, especially in the city of Dhaka. Unfortunately, the city government is not well organized to handle this task. Every day, about 3,500 – 4,500 metric tons of solid waste is generated in Dhaka of which only 50 per cent is collected by the Dhaka City Corporation (DCC), community-based Organizations (CBOs) and private organizations (POs). The remaining 50 per cent of the solid waste is discarded in streets, drains, ditches, canals and open spaces.

The population of Dhaka, the capital city, has grown significantly in recent years. A significant section of the population consists of rural migrants who are driven to the city in search of livelihoods. The deplorable condition of these poor migrants and the acute shortage of adequate housing has led to the growth of slums and squatter settlements in Dhaka. It is estimated that these settlements are inhabited by nearly one-third of the city's population. Municipal services are already overburdened. The rapid growth of the urban population also puts tremendous pressure on the performing agencies. They simply cannot provide adequate sanitary and other basic services to the slum-dwellers, resulting in unhygienic and filthy living conditions. DCC is also unable to provide all municipal services because of inadequate resources, limited institutional capacity, inefficient operational practices and corruption, including the lack of transparency.

In recent years, the role of public authorities in metropolitan management has been taken over by CBOs and POs in some parts of Dhaka city due to DCC's resource constraints and other limitations. These organizations are playing a partnership role in community management. The CBO approach is considered to be among the best in domestic solid waste management. 'Waste Concern', a CBO, has been playing a substantial role in implementing community waste management practices. The present paper discusses in detail the problems of waste management in Dhaka City and the role of DCC and some of the community based initiatives in solving the problems.

Urban Governance and the Role of Local Government: Bangladesh

Urban governance plays a key role in city management. However, in Bangladesh it is not properly organized even though Dhaka has a long checkered history.[1] Broadly speaking, urban governance refers to the system of government – concentrating on effective and accountable institutions, democratic principles and electoral process, representative and responsible structures of government – in order to ensure an open and legitimate relationship between civil society and the state (McCarney 1995). Governance is also defined as "the experience of authority, control, management, power of government" (World Bank 1992). In the context of governance, two aspects should be mainly considered, viz., political and administrative.

Municipal governance issues – including the interaction of central government institutions and the current structure of municipal government in relation to other levels of government – have a major impact on the effective provision of services by municipalities and on improved municipal financial management. Bangladesh already has the legal-institutional framework for a decentralized system of government, with democratically elected officials. The Government of Bangladesh has for some time emphasized the importance of devolution of power through local government institutions as a way of improving performance and fostering economic development at the local level.

Recently, the functions and responsibilities of public authorities in metropolitan management have begun to be shared by private organisations and Community based organizations in some parts of Dhaka city due to resource constraints and lack of state-driven initiatives in the operation and management of urban environmental issues. This has led to the formation of CBO and PO partnership initiatives in the sphere of municipal services and management. It is noted that decentralization, transparency and people's participation are key to good governance.[2]

Present Structure of Metro Governance in Dhaka

At present metropolitan governance[3] in Dhaka is broadly shared by three types of agencies – national, sectoral and local. Within the sectoral agencies, there are some special agencies for the city. Thus, a multiplicity of agencies exists to guide and control the planning and development of Dhaka and its metropolitan areas. Twenty-two ministries, out of a total of 38, and 51 agencies are involved in the planning, execution and development of metropolitan Dhaka.[4] The sectoral agencies are dominant and manage the better part of the city (66.7 per cent) as against the weak position of the urban local government, that is DCC which manages only 13.7 per cent of the city area.

There are structural relationships among various urban institutions that determine Dhaka metropoiltan development plan (DMDP). DMDP's planning and development is the primary responsibility of the institutions belonging to the Ministry of Housing and Public Works (MHPW). However, the Ministry of Local Government, Rural Development and Cooperatives (LGRDC), and other ministerial agencies representing government departments and statutory bodies also contribute to shaping the urban affairs of DMDP.

Generation of Waste

The quantity of solid waste generated in Dhaka city is increasing in proportion to the rate of population growth. The solid waste generation in the city varies between 0.47 kg/per capita / per day to 0.52 kg/cap/day (Table. 2 and 3 show the quantity and composition of solid waste).

Table. 1: Waste Generation in Dhaka City

Year	Waste generated (metric tons per day)
1990	2,000
1995	2,500
1997	3,000
2000	3,300 Projected *

*It is estimated that Dhaka City produced about 3,500 to 4,500 tons of waste per day.
Source: BCAS (1997).

Table 2: Composition of Different Kinds of Solid Waste in DCC Area

Component	Mixed Waste by weight(wt) (%)	Residential Waste by wt (%)	Commercial Waste by wt(%)	Industrial Waste by wt(%)
Food and vegetable waste	70.12	59.91	62.05	26.37
Paper products	4.29	11.21	6.28	7.59
Plastic, rubber and leather	4.71	17.67	4.62	6.01
Metals	0.13	0.15	0.28	-
Glass and ceramics	0.25	-	0.37	-
Wood	0.16	-	-	-
Garden wastes, Tree trimmings and straw	10.76	8.76	2.86	4.32
Cloth	4.57	-	18.93	46.2
Rocks, dirt and miscellaneous	5.01	2.30	4.62	9.49
Moisture content (%)	65	50	54	60

Source: GoB and World Bank (1998).

Table 2 shows the physical composition of solid waste generated in Dhaka City. A major portion (70 per cent to 80 per cent) of the solid waste is organic in nature and is generated from the residential, commercial and mixed areas of the City. The composition indicates the necessity for frequent collection and removal of the city's waste. It also points to the potential for recycling the organic waste into compost. There is a good market for compost, as a major portion of the land in greater Dhaka and its adjoining areas is used for agriculture. Thus, compost generated in urban areas may be used for rural agriculture. Table 3 gives a picture of waste generated in the administrative divisions of Bangladesh in 1991. It shows that the Dhaka division generated the highest amount of solid waste in that year.

Residential households contribute almost half of the total solid waste in Dhaka city. Commercial and industrial sources are the next important contributors. They contribute nearly 20 per cent each. Street sweepings are also a major source of solid waste, accounting for as much as 15 per cent. However, some studies have totally ignored this source (Sumon 2000). DCC estimated in 1999 that residential waste accounts for 49 per cent of the total solid waste; commercial waste 21 per cent, industrial waste 24 per cent and hospital waste 6 per cent. City waste is generally considered to be municipal solid waste. However, municipal solid waste is augmented by other hazardous and toxic waste, such as clinical and hospital wastes as well as tanning wastes.

Household solid waste

The major sources of household solid waste in Dhaka are residential areas, market places and commercial establishments. As indicated in Table.1, Dhaka City generates 3,500-4,500 tons of waste every day. DCC is responsible for the management of municipal waste in Dhaka city. It covers an area of about 1,528 sq km, with a population of about 9.3 million comprising of 1.66 million households. On an average, each household generates a little more than 1kg of waste (BCAS, 1997).

Table 3: Solid Waste Generation in Bangladesh (1991)

Administrative Divisions	Urban			Rural			Total
	Population	Waste Generated		Population	Waste Generated		Waste Generated
	Million	* kg/cap /day	Metric tons /day	Million	kg/cap /day	Metric Tons day	Metric tons day
Rajshahi	2.213	0.5	1107	25.287	0.15	3793	4900
Barisal	0.466	0.4	186	7.291	0.15	1094	1280
Khulna	1.609	0.5	805	11.634	0.15	1745	2550
Dhaka	5.966	0.5	2983	27.974	0.15	4196	7179
Chittagong	2.619	0.5	1310	19.250	0.15	2888	4198
Sylhet	0.255	0.4	102	6.892	0.15	1034	1136
Total	13.138		6493	98.328		14750	21243

* Per capita generation rate – (kg/cap/day).
Source: World Bank (1998).

Clinical waste

There are over 500 small and large clinics and hospitals in Dhaka City, based on an inventory by the Directorate of Health. The present average clinical waste generation in hospitals and nursing homes is calculated using 1kg/bed/day and an extra 200 kg a year for small clinics. The figure works out to around 20 tons of clinical waste generated daily in the city. About 20 per cent of the total waste is infectious and hazardous (BCAS 1997.).

Waste is collected from a small bowl (plastic or metal) or plastic bin, provided for each bed in a hospital or clinic, and put either in a large size plastic bag or bucket (plastic or metal). These plastic bags or waste from the buckets are then placed in a pushcart and carried to the nearest municipal bins for dumping. The municipal bins are placed either on the hospital premises or outside the hospital. Waste from operation theatres, laboratories and hospital kitchens is also dumped into the municipal bins in a similar fashion. The waste is then collected from municipal bins by DCC's trucks and carried to the landfill area for final dumping (Ibid. 1997).

But, there are no official guidelines for the disposal of clinical waste, which leads to environment pollution and deadly infections. The authorities concerned give low priority to the checking and monitoring of disposal of clinical waste containing hazardous materials.

Industrial waste

Industries are mainly situated near riversides in Dhaka, thereby polluting the rivers. Pollutants present in industrial effluents can be broadly categorized into:

- Biodegradable organic compounds (measured in terms of biochemical oxygen demand or BOD[5])
- Non-biodegradable or persistent organic compounds (measured in terms of chemical oxygen demand or COD)
- Heavy metals (e.g. Hg, Cr, Cd, Pb, Cu)
- Wide range of other dissolved substances and suspended solids

The tanneries, textiles, paper and pulp industries, food and fish industries, and distilleries mainly discharge biodegradable organic wastes. These undergo aerobic (i.e., using oxygen) decomposition in the receiving water body. As a result, dissolved oxygen in the water goes down, endangering fish and other aquatic species. Non-biodegradable organic wastes are mainly discharged by the pesticide, chemical and pharmaceutical industries.

It is reported that tanneries are releasing more than 20,000 cubic meters of highly toxic liquid waste round the clock into the drains and open spaces of Hazaribagh (south-west of Dhaka city), first flowing into the surface drains of the area and then spreading to the low-lying areas in Lalbagh, Rayer Bazar, Mohammadpur and Kalyanpur inside the flood protection embankment of Dhaka city. The poisonous waste thereafter finds its way into the low-lying cultivable lands along the embankment or through the sluice gates into the Buriganga River. The Buriganga, the ultimate destination, is virtually devoid of aquatic life – so much so that the fishing community, which used to depend for its livelihood on this river about fifty years ago, has now lost its means of subsistence.

Present Management System: Organizational and Institutional Aspects

Dhaka city suffers from poor solid waste management practices. The solid waste collection and transportation system has serious resource constraints and limitations – associated with planning, management, operation and maintenance.

According to the Dhaka City Corporation Ordinance of 1990, management of the solid waste of Dhaka City rests with the DCC. In the DCC such responsibility lies with the Conservancy Department. The Department has 5,200 conservancy and 135 supervisor staff for this purpose. The Chief Conservancy Officer is assisted by a Deputy Chief Conservancy Officer and two assistant conservancy officers. Ten conservancy officers, working in 10 zones with conservancy supervisory inspectors, conservancy inspectors and cleaners, also assist the assistant conservancy officers. The Conservancy Department supervises the garbage vehicles with support from the Transport Department. The cleaners of DCC sweep the roads and clean the drains daily. They accumulate the waste on the roads and along the drains. Once the street sweeping and drain cleaning have been done, cleaners collect the waste in cane baskets and handcarts for dumping at the nearest collection point or dustbins. City-dwellers dump their household waste at nearby DCC trucks for final disposal in the Muatail landfill areas.

The present legal foundations of solid waste management are the Municipal Ordinance 1983 and the Bangladesh Environment Protection Act (EPA) 1995. DCC, however, is responsible for collecting and cleaning about 80 per cent of this waste. The rest is to be disposed of by the Railway and the Public Works Department (*The Daily Star*, 1998).

But only 50 per cent of Dhaka City's garbage is estimated to be collected by the municipal corporation. The remaining half is discarded in streets, drains, ditches, canals and open spaces.

Therefore, large quantities of solid waste remain uncollected for long periods, leading to public health risks and environmental degradation. The crude dumping of solid waste at low-lying areas also poses serious health and environmental hazards, including air, surface as well as groundwater pollution. Lack of bins or their limited numbers in many city localities encourages people to throw their household waste on the streets and in open spaces. The gap between waste generation and disposal and collection by DCC results in overflowing bins/containers, with waste often being dumped around these bins/containers. This creates an unhygienic environment in the surrounding areas, including busy roads.

Garbage is dumped at the final dumping ground without taking any pollution control and treatment measure. This also leads to serious pollution of the surrounding environment, groundwater and soil and causes great harm to public health. Such crude dumping is regularly practiced. This is said to be due to lack of financial and technical resources. But despite 165 garbage-carrying trucks, 140 de-mountable trucks, 3,000 handcarts, 5,200 cleaners and a budget of over Tk. 95 million, DCC is providing an inadequate level of conservancy services. Over the years, DCC has increased its manpower and equipment, but there has been a little improvement in the quality of the services provided. Dhaka Municipality was established in 1864 and upgraded to a city corporation in 1990. But, since its inception only two significant developments in the area of solid waste management took place in DCC. In 1982, open trucks replaced bullock carts for collection and transportation of solid waste; and in 1989, night collection was introduced (Enayatullah 1995).

Drawbacks of the Present System

Some drawbacks of the present waste collection system of DCC are: rampant corruption, lack of accountability and poor supervision. About 50 per cent of DCC's conservancy manpower remains absent each day. Each sweeper has to clean only about 100 meters of roads, and each truck has to make a trip of 2 km a day (Assaduzzaman 1998). The DCC has to arrange for the sweeping of about 2,395 km of roads, streets and footpaths and the cleaning of about 2,463 km of drains every day. The 288 sq.km area of the city is divided into 10 zones for the purpose of collecting solid waste. The cleaners of DCC are supposed to sweep the roads and clean the drains daily in three shifts – morning, day and night. It is estimated that the support services, manpower and equipment that are currently available to DCC, if properly utilized, should be sufficient for ensuring solid waste collection and disposal. Although the DCC has increased its workforce and equipment for solid waste management and almost doubled its budgetary allocation for the purpose in the last four to five years, there has been little improvement in the quality of services. The cleaners hardly follow their schedule; as a result, garbage remains piled up. The DCC authorities, because of alleged pressure from workers' union and groups, have never effectively dealt with such gross neglect of duty. The system of reporting for duty is full of loopholes, breeding non-accountability (*The daily Star*, 1 August 1998).

To sum up, the following points may be considered as the main bottlenecks in solid waste disposal in Dhaka city:

- Negligence of duties and non-accountability
- Flawed reporting system and office hierarchy
- Lack of supervision
- Lack of coordination
- Lack of professional workers
- Lack of people's participation

Alternative Strategy for Municipal Solid Waste Management – Need for GO-NGO Partnership for Sustainable Solution

Due to the absence of regular municipal waste collection and the non-availability of municipal bins at suitable locations, waste is found scattered on the streets, presenting an eyesore and posing health risks. This state of affairs has forced local communities to establish their own informal system of waste collection. Community-managed house-to-house waste collection is gaining momentum and gradually expanding into a major environmental movement.

The DCC, in recent times has initiated some efforts to involve industries. In collaboration with the Bangladesh Council of Scientific and Industrial Research (BCSIR), DCC runs a pilot biogas plant using solid waste. In the DCC area, the Local Government Engineering Department (LGED) is also running a composting plant, which uses the organic matter in solid waste on a pilot basis.

In many areas of Dhaka city, like Kalabagan, Kathalbagan, Shamoly, Mohammadpur, Pallabi, Banani, Uttra, Gulshan and Baridhara, there are good examples of private initiative in the area of solid waste management. The primary objective of the private initiative is to provide a refuse-collection service from door to door so that the locality remains clean. Only kitchen garbage is collected from the participating households under these private schemes and is carried to the nearest DCC garbage bins or de-mountable containers for collection by the DCC trucks for final disposal.

Community mobilization is an important instrument in making the programme a viable one because people's participation is very uniquely required in the initiation and the sustenance of the programme..

Between DCC and small communities, NGOs can play an intermediate role in coordinating community participation. The system is encouraging as people are willing to pay for a cleaner neighborhood. If the programme is to be a sustainable one, it is necessary to introduce recycling. DCC has limited capacity, capabilities, financial resources and organizational structure. Consequently it has to look for an economically viable option for the management of waste through GO-NGO partnerships. A case study of Waste Concern in association with the Environmental Task Force of the Goethe Institute, Dhaka, is a landmark example. The partnership between DCC and Waste Concern (Figure. 1), a local NGO with the aim of converting waste into a resource that can be integrated with informal sector initiatives is a novel experience. To start with, DCC has agreed to set up a 1-ton capacity composting plant as a pilot programme to assess its technical and financial viability

Figure 1: Partnership Model.

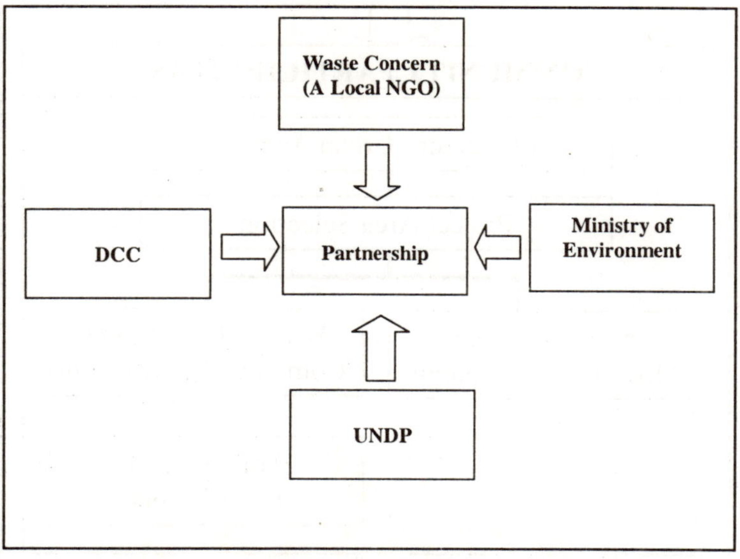

Source: Tariq, 2000.

Community-Based Decentralized Composting

To recover value from the organic portion of solid waste, Waste Concern initiated a community-based Decentralized Composting Project (Figure. 2) in Mirpur, Dhaka. A small piece of vacant land, measuring 1000 sq. mt, within Section-2 of Mirpur

Housing Estate was made available to Waste Concern by the Lions' Club (Dhaka North) for the composting plant.

The inspiration for the community-based waste management activity came from the basic outcome of the Kalababagan experience. Its results were positive – the locality became significantly cleaner and drains were no longer blocked with waste, leading to a reduction in waterlogging and the attendant problem of mosquitoes.

Figure 2: Three Major Components of the Decentralised Composting Project

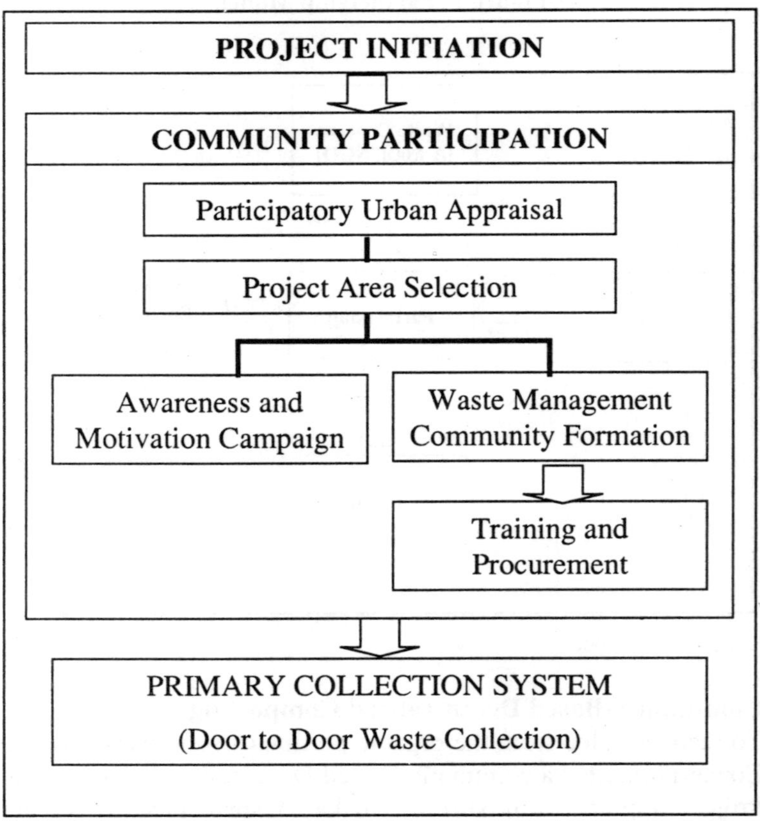

Waste Concern gave special attention to assessment of the needs and aspirations of the beneficiaries and to community involvement in the project. With this objective in mind and before initiating the project, Waste Concern, with the help of the Environmental Task Force of the Goethe Institute conducted a questionnaire survey among the residents of section-2, Mirpur Housing Estate. The purpose was to ascertain their opinion on solid waste management, their motivation to participate in any improvement programme and their willingness to contribute. The results of the questionnaire survey provided the basis for the project. The survey findings revealed that more than 80 per cent of the residents of the locality were not satisfied with the existing solid waste management service of DCC.

Most respondents supported the idea of an alternative door-to-door solid waste collection scheme. In addition, when the respondents were asked it they were in favour of paying a service charge for house-to-house waste collection, 77 per cent of the them expressed a willingness to pay Tk 15-60 per month to any organization for the additional service (Enayetuallh 2000). These findings encouraged Waste Concern to initiate the project. The remaining households were of the opinion that since they were paying conservancy tax to DCC, they were not prepared to pay an extra service charge for the door-to-door waste collection.

In addition to community mobilization, Waste Concern has also been working to create awareness among city-dwellers regarding source separation, recycling and resource recovery of solid waste. Towards this end, it has brought out posters and training manuals. In order to increase awareness, Waste Concern has established links with the residents' association of the neighbourhood.

A door-to-door solid waste collection system was introduced by Waste Concern to collect domestic organic waste (free from toxic and clinical waste) in the project area. At present, it collects 2 tons of solid waste per day. Modified rickshaw vans are used for collection. Before obtaining source-separated waste from the households, solid waste from restaurants and vegetable markets are collected separately. Initially, 100 households of the area participated in the programme. Four years after its launch 1,000 households of Section-2, Mirpur Housing Estate, are participating in the waste collection scheme. Of these, 700 households are participating in the door-to-door collection programme of Waste Concern and 300 households are participating in a similar programme run by another CBO. The door-to-door solid waste collection scheme of Waste Concern has had a demonstration effect. Four years after it began, adjoining neighbourhoods have also started similar schemes and formed CBOs. Waste Concern plans to expand its waste collection activities to 1,000 households and has also launched an environmental awareness programme with the help of the community.

About one-third of Dhaka's population lives in slums and squatters. Therefore, to develop an alternative low-cost and sustainable solid waste management option for slums, Waste Concern, for the first time in Mirpur initiated a pilot barrel type composting project in two slums, with support from UNDP's Local Initiative Facility for Environment (LIFE) programme. The programme was replicated in other parts of the city owing to its success and with the aim of bringing sustainable solid waste management services with economic benefits within the reach of slum-dwellers. There is a good market for compost around Dhaka. Waste Concern sells its compost to a number of outlets, like fertilizer marketing companies and nurseries.

Conclusion and Recommendations

Solid waste management in Dhaka City suffers from several drawbacks. The roads, streets and drains remain unclean and dirty, creating serious health and environmental problems. The DCC is unable to tackle them alone. The present system is based on crude dumping which is not sustainable from an environmental point of view. Almost 90 per cent of the solid waste generated in the DCC area is recyclable. It is, therefore, necessary that recycling of waste is given priority.

The DCC, with technical expertise from BCSIR and LGED, should construct the biogas and composting plants in different parts of the city. The size and location of these plants should be determined by the amount and rate of waste generated in the area. The personnel of the DCC, BCSIR and LGED should be initially responsible for the operation and maintenance of the plants. The Department of Environment would supervise the functions of the plants. There is an absence of skilled manpower in DCC to run the plants. BCSIR and LGED could help in training the DCC personnel. Since the government is under pressure from donor agencies to downsize, workers of CBOs should also be trained to run and maintain the biogas and composting plants. Thus, if the government plans to reduce the staff of DCC, there will be back-up team for taking over the recycling plants.

The emergence and participation of CBO's is an important development in the search for solutions to the massive problem of solid waste disposal. The recent experience of Waste Concern in decentralized composting in the city of Dhaka has been promising and encouraging. The results from the five years of operation of the pilot project support the hypothesis that small-scale composting is commercially profitable, mainly because the method adopted produces high-quality compost from raw waste.

For the economic viability of collaboration between CBOs/NGOs and local government agencies for a cleaner community environment and prevention of the spread of infectious diseases, there should be adequate financial allocation for SWM. Adequate finance could be attained by the following measures:

- Conservancy tax should be delinked from general property tax based on rental value. It is recommended that the market value of property be used.
- Separate rates should be worked out for collecting commercial (including retail shops) and industrial wastes.
- Charges should be levied for removal of construction or demolition waste. This tax can be levied while granting building permission.

- The success of a community-based programme depends largely on identifying and addressing the community needs; while the sustainability of the project depends on involving them in the cost-recovery/cost-sharing process.
- Community-based projects have demonstration effects, as proved by the Mirpur project. Such projects can be replicated in other parts of the city.
- NGOs can play an important role in initiating and demonstrating new concepts and providing technical know-how and training to the community participants.
- Small-scale compost plants can be located within the community via joint collaboration between the DCC and CBOs/NGOs, provided an appropriate and scientific composting method is followed.

Despite the achievements of CBOs/NGOs, the programme has been slow to expand. Replication of the concept in other communities, city corporations and municipalities has been rather unsatisfactory. At the national level, there needs to be a policy in Bangladesh on solid waste management for city corporations and municipalities.

NOTES

1 Dhaka is an old town and is known to be in existence since the seventh century AD Islam Khan, the first Mughal Viceroy of Bengal, founded Dhaka as the capital of Bengal (Mohit 1998).

2. Good governance means less direction and control and more freedom of action (Muhit 2001).

3. There is no structure for metro governmance in Dhaka. The city also faces a lack of coordination between the national, sectoral and local agencies.

4. Metropolitan Dhaka has two connotations. It refers to the central city – Dhaka City Corporation (DCC) or Dhaka City, as it is popularly known; as also to the Dhaka Statistical Metropolitan Area (DSMA), which includes DCC as well as Keranigonj, Demra, Narayangonj, Dhaka Cantonment, Tongi, and adjoining rural areas. The total area covered by

DCC is only about 200 square miles. The DSMA area is about 550 square miles. The Dhaka Metropolitan Development Plan (1995-2015), prepared by RAJUK (Capital Development Authority), contains an area of 590 square miles (1,528 square kilometers).

5 The amount of oxygen used by microorganisms in the process of breaking down organic matter in water. More the quantity of organic matter (e.g., in sewage), the greater is the number of microbes. The greater microbes again lead to the greater need for oxygen to support them; consequently, less oxygen is available for fish and other animals in water. BOD is, therefore, a reliable gauge of the organic pollution of a water body. One of the main reasons for treating sewage or waste water prior to its return to a water resource is to lower its BOD; and thereby lessen the demand for oxygen from the streams or rivers into which it is released.

REFERENCES

Amin, A T T N 2000, *Transforming Waste into Resource as a Basis of Building a New Urban Community: An Overview of Regional Experience*, Waste Concern, Dhaka.

BCAS 1997, " Waste Management in Dhaka City", *Bangladesh Environmental Newsletter*, vol. 8 no. 4, October-December 1997.

Enayetullah I, A H. Sinha and Md. Maqsood 2000, *Barrel Type Composting for Slums: Experience of Waste Concern in Dhaka,* Waste Concern, Dhaka.

Enayetullah I 1995, *A Study of Solid Waste Management for Environmental Improvement of Dhaka City*, unpublished MURP thesis, BUET, Dhaka.

McCarney P L 1995, "Towards an Understanding of Governance: The Emergence of an Idea and Its Implication for Urban Research in Developing Countries", in R Stren and J K Bell (eds.), *Urban Research in the Developing Countries, Vol.4, Perspective on the City,* Centre for Urban and Community Studies, University of Toronto, Toronto.

Mohit A Mohammad 1998, *Dhaka Planning and Development Problems*, Department of Urban and Regional Planning, Bangladesh University of Engineering and Technology, Dhaka.

Muhit A M A 2001, "Issues of Governance in Bangladesh", *The Daily Star*, 24 June, Dhaka.

Sumon Md. Abu 2000, *Developing a Sustainable Solid Waste Management System as an Alternative Option for Urban Communities*, Khulna University, Khulna, Bangladesh.

Tariq Yousuf Bin 2000, "Community Waste Management: Possibilities and Partnership", *Community-Based Solid Waste Management: The Asian Experience,* Waste Concern, Dhaka.

World Bank 1992, *Governance and Development*, World Bank, Washington DC.

16

Creating Space for the Poor in Bankable Urban Infrastructure Projects

Renu Khosla & Chetan Vaidya

Introduction

Under the Indo-US FIRE project, it is planned to improve access of the poor to urban basic services as part of bankable water supply and sewerage projects in the cities of Ludhiana in Punjab and Sangali in Mahrashtra. The objective of the programme is to mobilize demand and institutionalize procedures for low-income settlements to access basic services on a sustainable basis. The approach to be used in the two cities is based on the experiences of community mobilization under several successful initiatives in India, particularly the Primary Education Enhancement project. The paper presents a road map for city managers to guide the process of community participation and planning for the poor. Drawing from the lessons of other social development initiatives, this paper has made some recommendations relating to the establishment of a policy framework within which city-level initiatives for improving access of poor to basic services must be taken.

Bankable Infrastructure Projects: Improving Access for the Urban Poor

The goal of the Indo-USAID Financial Institutions Reform and Expansion (FIRE) Project is to improve the delivery of environmental infrastructure in cities, in particular, access to basic services for urban poor households in bankable infrastructure projects.

Bankable or commercially viable urban infrastructure projects are those which are able to raise resources from the capital and financial markets, largely on the basis of revenue streams from specific service-linked user charges and dedicated sources, such as property tax and octroi. Commercial viability requires that there is (*a*) an adequate rate of return on investment; (*b*) appropriate institutional structure; and (*c*) a risk assessment and mitigation plan.

While developing plans for urban infrastructure development, specifically as part of commercially viable projects, needs of the poor are not factored in due to their low paying capacity and welfare orientation of the city plans. On the contrary, sufficient evidence exists with regard to the willingness of the poor to pay for good quality services (NIUA 2001). Not only are the poor disposed towards paying for access to water and sanitation services, they are currently doing so, albeit to the local mafia.

Bankable urban infrastructure projects can simultaneously be poor-friendly and meet their obligations vis-à-vis cost recovery. The FIRE project plans to demonstrate this strategy by developing an action plan to improve access of the poor to urban basic services as part of water supply and sewerage projects in two cities – Ludhiana in Punjab and Sangali in Maharashtra. This is a major partnership programme of the urban local bodies (ULB), Indo-USAID FIRE Project, the National Institute of Urban Affairs (NIUA) and the Cities Alliance.

The objective of the programme in the two cities is to mobilize demand and institutionalize procedures for low-income settlements to access basic services from ULBs on a sustainable basis. The endeavour is to address the needs of the poor on a city-wide basis, as also to coalesce the efforts of the multiple urban partners towards achieving the vision of cities without slums.

Opportunities in the Two Cities: Connecting Big City Pipes to Small Household Pipes

To understand the context within which the programme was to be located and the reasons for targeting the two cities, a rapid situational analysis was made, which is presented below.

Ludhiana

Ludhiana, an industrial hub of Punjab, has an estimated population of 1.4 million as per Census 2001. It has attracted people from across the country because of its huge employment potential. A majority of these are migrant workers who live in underserviced settlements of the city due to their inability to afford better housing.

With committed and visionary leadership, the Municipal Corporation of Ludhiana (MCL) has introduced a number of innovations towards good city governance. Key among these include, improved financial performance and issuance of municipal bonds, community participation in improving sanitation and maintenance of neighbourhood parks, private sector participation (PSP) in water bill distribution, outsourcing of solid waste transportation, and disposal and introduction of a double entry accrual-based municipal accounting system.

MCL is implementing a Rs 330 million water supply project and a Rs 1,500 million sewerage project with the objective of ensuring universal coverage of the city. The projects have been financed through borrowings from HUDCO. These ongoing water supply and sewerage projects are expected to bring trunk infrastructure close to most low-income areas in the city. With a little effort, the poor settlements in the city can be linked to city-level water supply, sanitation and solid waste collection services. Household services can be made available to the poor in these settlements by organizing them and enabling them to identify and prioritize their needs and to mobilize resources for extending small pipe connections from larger city pipes to their homes. Simultaneously, the development of performance criteria could enable city governments to monitor the availability of these services and their access by the poor.

The task for Ludhiana is enormous and complex in view of the magnitude of the poor population in the city. Its 182 low-income settlements with around 60,000 households are largely situated on private lands. Nine recognized slums are on government land and are being serviced albeit through common community services such as public stand posts and toilets. Some settlements on central government lands, such as railways, exist but have yet to be listed by the MCL and hence remain outside the purview of service provision. The process of community participation in improving access to basic services has yet to be institutionalized. Some effort at building community organisations, though, is visible under the national poverty alleviation programme and the Swarna Jayanti Shahari Rojgar Yojana (SJSRY), but the focus is on provision of loans for income generation. Under the SJSRY initiative, Community Development Societies (CDS) have been formed in about 52 settlements, some of who are successfully engaged in economic activities. CDS groups are also expected to prepare mini plans that are to be integrated into a pro-poor city plan of action. This, however, has not happened.

Box No. 1: Swarna Jayanti Shahari Rojgar Yojana

A national urban poverty alleviation programme, the Swarna Jayanti Shahari Rojgar Yojana (SJSRY) was started in December 1997 to reduce income-based poverty in cities. Its primary objective is to make available credit assistance to the poor for setting up individual enterprises and to provide opportunities for building the capacities of the poor to engage in skilled employment. The credit assistance programme was linked with a strategy of organizing the community into neighbourhood groups, neighbourhood committees and community development societies, as recommended under the Urban Basic Services for the Poor (UBSP) Programme. Its two major components are the Urban Self-Employment Programme (USEP) and the Urban Wage Employment Programme (UWEP).

Shangli

The Sangli-Miraj-Kupwad Municipal Corporation (SMKMC) in Maharashtra was formed in 1998 with the merger of three neighbourhood municipal councils and is spread over an area of 118 sq. km. The combined population of SMKMC was estimated at 4,50,000 in 2000. According to SMKMC, the city had about 60 low-income settlements, housing nearly 6,500 families. Of these, the 28 recognized settlements have 4,572 households and an average household size of 6.6. A recent survey initiated under the FIRE project has, however, identified 35 new low-income settlements, taking the total number of poor settlements in the city to 95. Most low-income settlements in the city are located on public land, except for nine that are on private lands.

Innovations at the local level in the past three years have included the passing of a resolution by the SMKMC, followed by the signing of a memorandum of agreement with the Infrastructure Leasing and Financial Services Limited (IL&FS)

to form a special purpose vehicle for private sector participation in the city's water supply and sewerage systems. SMKMC proposes to issue a three-year management contract to a private operator to maintain the existing water supply system, reduce unaccounted for water, save energy and improve billing collection.

Other innovations being instituted by SMCMC are: preparation of a development plan through private consultants, solid waste collection through private service contracts; computerization of accounts and records of the local body, networking of all zonal offices in the city through local area network; introduction of area-based property tax assessment for new properties; and a double entry accrual-based municipal accounting system.

In keeping with the 74th Constitutional Amendment Act (CAA) 1992, SMKMC has decentralized local governance by setting up four wards committees, each comprising 15-17 wards. Wards committees have been given both functional and financial autonomy to take decisions for the development of their areas. Funds have been pooled under several programmes, including the National Slum Development Programme (NSDP), SJSRY and councillor's funds. They have been augmented by settling aside 5 per cent of the net municipal revenue for providing services in low-income settlements.

Proposed Interventions in Ludhiana and Sangali under the FIRE Project

A three-track approach is planned to network poor settlements to the city within the framework of commercially viable projects.

A. Improve basic service provision through mobilization and organization of communities in underserved settlements, including their economic empowerment, awareness building and establishment of linkages with ULBs and other government departments.

B. Support the development of processes and procedures in the ULBs and build staff capacity and the commitment to respond to demands for service provision in the presently underserved communities.

C. Encourage and facilitate convergence of resources under various relevant state and central government programmes and cost contributions from communities to finance the required infrastructure provision in underserved settlements and address critical issues such as land tenure and decentralized management.

Community Building Processes

In order to achieve the aforementioned objectives and bring people's concern centre stage, a process of community mobilization is proposed.

The community mobilization experience in Delhi, facilitated by NIUA under the Primary Education Enhancement Project (PEEP), serves as an excellent model for initiating and scaling up the current community organization processes underway in the identified cities, building local support capacity and enhancing dialogue with the authorities.

PEEP – a partnership between all the principal stakeholders in the education department in the city, UNICEF, AUSAID, NGOs and NIUA -- aims at universal coverage of all children, particularly from disadvantaged communities, for education in association with communities. Community building processes underway in over 300 squatter settlements in Delhi include:

- Organizing women into neighbourhood groups and *bastee* (slum) committees
- Empowering the poor to identify their needs and priorities, using a range of participatory instruments
- Facilitating the process of planning action at the community level to address needs
- Mobilizing resources for meeting immediate priorities
- Enabling the poor to negotiate with the government for their entitlements

Although the process of community building under PEEP centres on education, needs of the poor have been found to range from basic survival (water, sanitation and health) to improved quality of life through individual services and greater representation at the local level in decision making.

The capacity of women community representatives is enhanced through training programmes that are based on their identified needs and priorities. Networking of community groups at the ward and zone level is facilitated to improve their visibility and audibility at the city level. Community groups have addressed their need for better quality education through the establishment of sahayak shiksha kendras or education support centres that are self-initiated and self-managed to provide after-school help or early childhood stimulation to children.

Women's groups, besides attending to their education concerns, have also engaged with area-level officers and elected representatives for the provision of basic infrastructure in these settlements. They have shared the responsibility for cleaning up public toilets in their blocks; and through a user charge, are able to maintain them in a usable condition. They have ensured regular water supply through tankers to community doorsteps. The tension between upper-income communities and slum neighbourhoods over the passage of these tankers was overcome through a negotiated settlement that led to mutual gains. The construction of a community centre within the premises of a neighbourhood temple was undertaken in partnership with a private operator, with the bastee committee taking the lead in bargaining with the temple authorities for space. Women's groups have picketed the area offices of sanitary inspectors and electricity officials to ensure services in their areas as per demand. Through voluntary work or *shram dan,* communities have helped to improve the infrastructure of the neighbourhood school. Self-help groups are being formed in many communities to provide emergency credit to women members and opportunities for saving up. Women have imbibed skills to negotiate with local governments for their entitlements.

Women in these communities have developed maps of their areas, indicating the availability of services in the settlements and have discussed infrastructure problems and their solutions. These maps have been built into a Community Based Geographical Information System (CB-GIS), which provides information at the city, community and household level with regard to availability of services. Using spatial tools, the CB-

GIS helps planners to examine issues of access and availability and to plan intervention based on an incremental model towards community upgrading.

Pro-Poor Focus in Bankable Urban Infrastructure Projects

Based on the community organization strategy demonstrated under PEEP, communities in Ludhiana and Sangli are to be organized for infrastructure development. Identification of the existing level of services in all the poor settlements of the city through community mapping would become the basis for preparing local area action plans by people centred on their identification of concerns. Facilitated by NIUA and local partners, a 'bottom-top' planning process would enable the integration of community plans into city infrastructure plans with clearly identified contributions by people.

A key problem in addressing the needs of the urban poor on a city-wide basis results from the lack of basic data about the nature and magnitude of the problem. As a first step, therefore, a database of low-income settlements (recognized/unrecognized slums and unauthorized colonies) will be created, indicating the number, level of services, location and ownership status of settlements. Ward, zone and city maps will indicate the access of low-income settlements to services of education, health, transport, food grain from ration shops, electricity, built roads, etc.

Information will be gathered using participatory instruments at the settlement level, such as ownership of house, years of stay in the settlement, number of households and population, current level of community /individual services for taps and toilets, municipal waste collection bins, street lights, drainage (covered/open), etc. Information on the problems of people vis-à-vis service delivery will be integrated with city-level data to help develop a spatial information system that provides community-based information at the city level. Individual settlement maps showing linked roads and infrastructure channels will also be developed to identify networking solutions for upgrading.

The process of community mobilization will be facilitated with the help of local NGOs/CBOs, who over the project phase will be associated with an NGO network with enhanced capacities to work in partnership with the local governments. The socio-economic and health profiles of communities will be benchmarked to enable impact assessment, both immediate and in the long term. Solutions at the area level will rest on the principles of ownership, self-reliance and choice, with communities mobilizing resources for household connections and agreeing to pay user charges for better quality services. Community contributions/commitment towards infrastructure building and payment of user charges will ensure that the intervention projects become commercially viable. Towards this end, communities will be helped to set up thrift and credit groups and enabled to access micro-credit for household investments on easy interest rates.

All key stakeholders in city development, including the private sector and other non-conventional actors, will be involved in the development of area plans and the identification of resources/opportunities at the city level. Systems of interaction between community associations, municipal officials and other stakeholders/partners will be developed to ensure sustainability of the endeavour.

While the approach will be to universalize coverage in the city, the process will be scaled up gradually after demonstrating its potential in a few selected settlements where linkages with the trunk infrastructure is possible. Efforts will also be made to put in place systems for community management of operation and maintenance(O& M) of basic services.

A pro-poor city-level strategy and action plan will be developed with a view to improving the access of urban poor to basic services, in particular sanitation. This will be done by integrating community action plans within city plans and water and sewerage plans. As part of the city plan, constraints in addressing the needs of non-notified communities, communities on private land and the poor in scattered settlements for basic infrastructure will be identified and alternatives for provision of services explored. Capital investments and O&M costs required for universal outreach of basic services will be envisaged and included. Project capacities would be planned keeping in mind the expansion of the city and the rate of migration into the city.

Implementation of the pro-poor city action plan will be supported by a demand-based capacity building programme for the ULB staff. This will enable them to work with poor communities; build community organizations and assess the level of community participation; use information systems for improving the governance of poverty alleviation strategies at the city level; and promote inter-sectoral convergence. In order to monitor the intervention and its impact, appropriate contract and performance indicators will be developed for the proposed private operator.

Cities without Slums within Bankable Projects: Suggestions

The poor are an essential part of a city, with their enormous contribution to city services and informal sector participation. In spite of their very significant role in the functioning of cities, they remain invisible and inaudible within the city plans for service delivery. Successfully integrating them with existing city systems will, therefore, depend on several initiatives at the city level.

1. Providing Legal Tenure

A city's inability to provide services to the poor is directly linked to the legality of the sites on which they squat. Illegal tenure deems the poor to be illegitimate citizens without rights to the city's services. Marginalization of the poor for service delivery, therefore, pushes them into living in unsanitary conditions in underserviced areas, making them highly vulnerable to their environments. Acknowledging the existence of the poor in unauthorized settlements and granting them tenure would be the first step towards making cities slum free. Tenure rights have been known to result in increased personal investments in shelter upgradation, particularly as large segments of the households in these settlements are not necessarily income poor, as was noted in the slum networking programme supported by the UK government's Department for International Development (DFID) in Indore.

Providing legal tenure will require revisiting the municipal act of the city and its provisions with regard to squatter settlements. Several examples are available across the country where *pattas or* land ownership titles have been granted to the poor, as in Madhya Pradesh, and fixed tenure has been provided -- in Ahmedabad under the amended municipal act.

2. Imposing User Charges

Levying user charges is an issue that has political ramifications. Often, elected representatives are unwilling to charge the poor for basic services even though the poor have demonstrated their willingness to pay for better quality individual services. Free/subsidized services result in poor financial health of the local body, which prevents them from making capital investments in the city's infrastructure or borrowing for the same from infrastructure leasing companies due to their low absorption capacity. Cities need to revisit their policy of free services to the poor. As a policy, they should be able to provide individual connections to the poor on payment of connection charges and monthly tariffs. Even community services may be provided at subsidized rates to increase community ownership of the services and, in turn, their sustainability.

3. Providing Access to Credit

To enable the poor to pay the connection costs, they may be provided access to loans at low- interest rates or helped to form thrift and credit groups. Tenure provision when linked to access to finance for shelter upgradation will enable the poor to borrow money for improving the quality of their shelter. Access to credit must be linked to employment generation activities. This will improve the absorption capacity of the poor to take loans and to repay them. Credit delivered through local community-based organizations will ensure pressure on the individuals to repay the loans. Promoting pro-poor small-scale lending schemes, low-cost technologies and access to markets/information will enable the poor to borrow money for improved services.

4. Community Management of Basic Services

Unusable community toilets, broken taps with treated water flowing down the drains, choked open drains and damaged paved pathways inside the settlement are common sights in low-income communities. They are a reflection of the lack of ownership and community involvement in the operation and management of services. The formation of cooperatives/management committees can enhance people's sense of ownership and responsibility in the maintenance of community facilities. City governments will need to formulate a clear-cut strategy for O&M of basic services within the overall planning for infrastructure delivery.

5. Community Contracting of Infrastructure Works Within Settlements

The promotion of community contracting for the development of appropriate urban infrastructure and services will ensure that every rupee goes further. Contracting out tasks of community building, such as road making or building community toilets, to community development societies (CDSs) has resulted in better quality of infrastructure, as in Andhra Pradesh under the NSDP and in Pune for toilet building. Community-based organizations (CBOs) are able to closely supervise the process of construction, ensure use of better quality material, as also cut costs by identifying local contractors/ labour at lower rates. SPARC's (Society for the Promotion of Area Resource Centres) experience in toilet building in Pune has shown that the overall product is far superior (Burra 2001).

6. Planning for Capital Investment: Cities must have a Vision

It is imperative that city governments understand the magnitude and multiple dimensions of poverty while planning for poverty reduction. Databases and information systems that provide current information and future projections are necessary for deciding on the engineering aspects of infrastructure building and the capacity of trunk infrastructure that needs to be augmented/built.

7. Gender-Sensitive Community Infrastructure Planning

Infrastructure must be gender friendly. Since conventionally, women are responsible for household chores, they are more influenced by the lack of basic services in the communities. Women are affected most when water taps dry up. In the absence of toilets, women are often subject to sexual harassment or have to walk under the cover of darkness in case of open defecation. Gender-friendly basic infrastructure will not only ease a woman's work burden but also increase her productivity by making available more time for economic activities. Services that meet the requirements of women also help to increase their sense of safety and well-being in the community. Promoting the participation of women in infrastructure planning is important to ensure that services are responsive to their needs.

8. Flexible Strategies

Cities need to evolve their own criteria for identification of the poor. Centralized norms based on incomes are proving inadequate for recognizing other vulnerable groups, such as the transitional poor (those with incomes just above the poverty line are vulnerable to economic shocks), and the core poor (pavement-dwellers and homeless people). The need for basic services varies according to the nature of the vulnerability. City plans must be flexible so that individual needs can be addressed. The development plans must allow time for nurturing community-building processes.

9. Process Development Support

The process of developing pro-poor urban projects in a commercial format is a relatively new trend in India. Unlike conventional projects, this type of process development requires considerable effort in terms of information collection, community participation, putting in place institutional arrangements, getting the approval of all the stakeholders, and management of project implementation. This process is time consuming and costly. ULBs often do not have the necessary human resources to carry out these tasks. International agencies and governments organizations need to support the process development activities.

Cities can be built without slums -- not by pushing the poor to the cities' fringes and beyond, but by recognizing that they have the right to live in cities and to use the basic services. Cities without slums are ones that integrate the poor within the urban fabric, such that they continue to contribute to the cities' economic development in a productive way. Local governments must acknowledge that the poor can afford to, and do already pay to access the basic services in cities and that it is possible to mobilize the community's own resources towards providing them better services. City governments can hope to develop commercially viable projects that address the needs of all the poor sections only when they are politically and administratively committed to ensuring the poor their rights to a better quality of life.

REFERENECS

Burra S, 2001, Slum Sanitation in Pune, Paper presented at the Good Governance Campaign, India Launch, UNCHS and Government of India, 4- 6 September 2001.

NIUA (National Institute of Urban Affairs) 2001, *Status of Urban Poor in Nagpur: A Benchmark Study,* NIUA New Delhi.

Parikh H, 1999, Pro-Poor City Planning: Using Slums to Save Cities, Paper presented at the Seminar on Social Dimensions of Urban Poverty in India, 3-5 March 1999, NIUA, India Habitat Centre, New Delhi.

17

Management of Urban Environmental Services

A Case Study of the Residential Sector in Vasai-Virar, Mumbai

Sumona Bandyopadhyay

Introduction

In the past few decades, the developing world has witnessed unabated population growth. While it is agreed that a manageable increase in population is desirable in an expanding economy, a crisis situation develops when industrializing countries witness population growth far in excess of their resource base.

The present urban crisis facing the metropolitan regions of India is a result of such uncontrolled growth vis-à-vis a fast dwindling resource base. Though ascertaining the optimum population size is a complex issue, the prevailing environmental problems are clearly discernible. Urban land and basic environmental services are the prime causalities in conditions of rapid demographic expansion. Of the major land uses in metropolitan areas, land set aside for residential purposes occupies the largest proportion; still, residential space is becoming a scarce commodity. Inadequate supply in the face of high demand pushes up real estate prices; this compels land developers to maximize space use, often at the cost of environmental health.

The present study examines the case of Vasai-Virar sub-region (VVSR)[1], a suburban residential sector in Mumbai. Here, the demand for housing and the related building activity is leading to the erosion of green belts, causing irreparable damage. Moreover, residential units are being constructed and handed over on ownership basis without having created the basic infrastructural services. The rationale behind selection of this area lies in the rampant unplanned development of this region, which has been categorized as a "green zone,"[2] primarily because of its ecologically sensitive character. The green zone of

Vasai-Virar, which was delineated by the development authorities, was intended to be a region where no urbanization would take place. It was supposed to be maintained in its natural form to serve two basic goals: first, to act as green lungs for the sustenance of the overpopulated and heavily polluted inner metropolis; second, to conserve the wetland region comprising numerous water bodies, channels and creeks. Urbanization of this green zone is characteristic of the prevailing apathy towards environmental considerations, which could prove to be disastrous in the long run.

The study is based on a survey conducted in 1997 and 2000 by the Department of Humanities & Social Sciences at the Indian Institute of Technology, Mumbai. The first phase of the study in 1997 was in fulfillment of the requirements of the M.Phil dissertation dealing with the analysis of environmental problems arising out of uncontrolled development and land use changes for the residential sector of suburban Mumbai. The sample consisted of 200 residents selected from Vasai, Virar and Nalasopara, to whom a questionnaire was administered. Another set of questionnaires were administerd to a select number of real estate developers and development authorities – City and Industrial Development Corporation (CIDCO) and Mumbai Metropolitan Regional Development Authority (MMRDA). Secondary data was collected from the MMRDA and CIDCO reports and documents.

Vasai-Virar Subregion (VVSR): Geographical Attributes

Vasai-Virar is located in the coastal zone north of Mumbai, demarcated by the Arabian Sea to its west and the Tungari Hills (an offshoot of the Western Ghats[3]) to the east. Vaitarna River forms its northern boundary while the Vasai Creek lies to its south. There are large tracts of wetlands within the subregion, and its proximity to the creeks and the coast gives it a unique ecological character. The wetlands of Vasai-Virar are microniches for many species of flora and fauna. (The white-breasted kingfisher is threatened by the damage caused to these wetlands.) Climatically, this microregion is characterized by uniformity in temperature and humidity. Summer temperatures range between 20 and 35^{0}C while the winter range is 16-28 ^{0} C.

The average rainfall is about 2,200 mm. The monsoon is the most striking climatic feature of this coastal region and extends from June to September.

Population growth in this region has been quite remarkable. The total population of Vasai taluka[4] alone was 1.22 lakhs in 1951; it had increased to 4.23 lakhs by 1991 (See Figure. 1). Figure. 2 shows urban population projections to be of the order of 1.7 million by 2011, while the total population of VVSR is projected to be 1.9 million. This implies that 92 per cent of the population of the region will be urban. The 1991 Census figures show that 83.4 per cent of the total population of VVSR is urban.

Figure 1

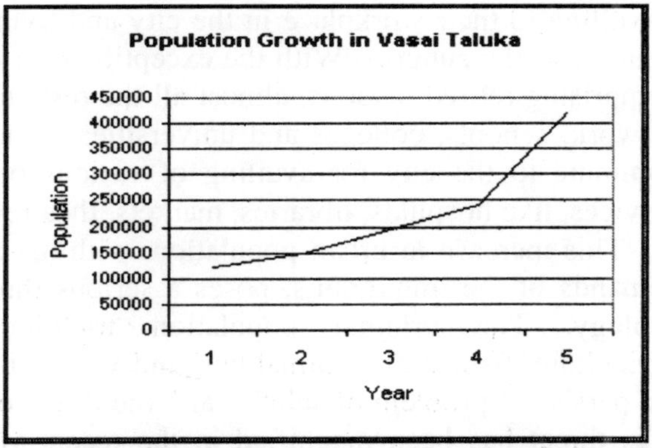

Figure 2

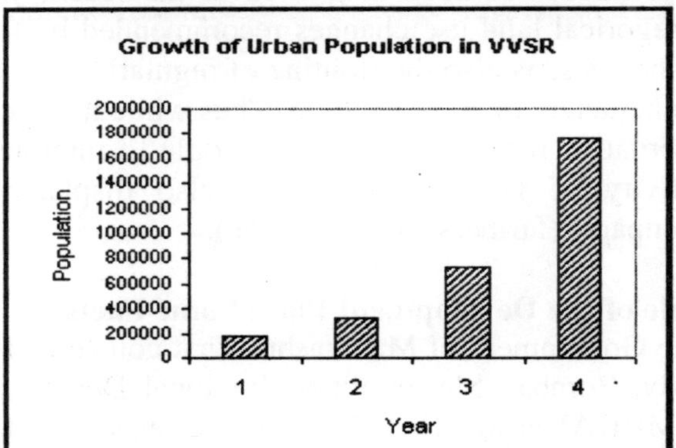

Note: Number on X axis corresponds as follows: 1=1951, 2=1961, 3=1971, 4=1981, 5=1991.

Source: Draft Development Plan for VVSR, CIDCO, Mumbai.

The area was predominantly agricultural until the development of cheap housing units started in late 1970s. As land prices skyrocketed in the inner city areas of Greater Bombay (now called Brihan Mumbai), this region became an affordable option and low-income group households started migrating into VVSR. Through the 1980s and early 1990s, not only low-income groups but also middle- and higher-income groups found it more affordable, especially as an option after retirement from service (this information is based on data collected from the sample on migration, place of origin and reason for migration). In the 1990s, the increased efficiency of the rail transit lead to an increase in the number of commuters travelling to their workplace in the city and living in affordable housing in the suburbs. With the exception of a few households comprising retired persons, almost all the respondents commute to work, schools, colleges and universities. The residents also commute to the city for availing of various other specialized services, like hospitals, libraries, markets, theatres, etc.

This increase in urban population, with the attendant rising demands of the immigrants, poses a serious threat to the local ecology. The original population traditionally practised agriculture, fishing and animal husbandry. But it was faced with the persistent problem of salinity and the threat of development. This threat has been magnified in the recent years due to the categorical land-use changes recommended by the development authorities, as also the flouting of regulations by builders, which is characteristic of this region. Thus while it is a cheaper housing alternative that is encouraging people to migrate to VVSR, the activity of land developers is also displacing the original occupants (farmers and fishermen).

Role of the Development Plan: Some Facts

The Government of Maharashtra first constituted a development body, Bombay Metropolitan Regional Development Authority (BMRDA) in March 1975, with the objective of ensuring three important aims:

1. Orderly regional development
2. Protection of green zones

3. Decongestion of the main city

According to the initial plan for the region, the Vasai-Nalasopara-Virar area, a primarily coastal region dotted with hills, was to be largely retained under the green zone.

Till 1975, the region escaped the attention of the builders because of the lack of infrastructure and the absence of potable water. In 1976, the Urban Land Ceiling Act[5] came into force but most of the Vasai-Virar area fell outside the purview of the Act. This attracted builders to the area. Simultaneously, the regulations also underwent some changes. Initially, only 776 hectares (2 per cent of VVSR) along the railway lines were to be urbanized. With growing population pressures on the main city, the government decided to increase this urbanizable area (termed urbanizable zone or U zone[6]) to 6 per cent in 1983. The urbanizable area, however, was not guided by any planning regulations, leading to haphazard growth. This did not elicit any response from the development and planning authorities. Instead, in 1988, the sanctioned regional plan was modified to convert 8,500 hectares of the green zone (G zone) into an urbanizable zone. In 1990, 2,000 hectares were further dereserved. This categorical dereservation of the green zone prompted several conscious citizens, along with some voluntary organizations, to file five writ petitions before the Bombay High Court. In a significant move, the court stayed all development activity in the 10,500 hectares of land dereserved after 1988 and the City and Industrial Development Corporation (CIDCO) was permitted to prepare plans for the area made urbanizable before 31 August 1988.

Both secondary sources and data collected from real estate developers indicate that despite legislation, construction activity continued to be reported and gram panchayats[7] continued to grant "no objection" certificates. Secondary sources reveal that 20,000 apartments were completed after the court order and another 4,311 were under construction even in June 1991. In the absence of infrastructure, the occupancy rates were low during this period. But by 1997-98, when this survey was conducted, these residential units showed full occupancy and construction activity was still continuing. The question that needs to be answered is whether sustainable solutions can be evolved under such circumstances.

Land Use Changes: Postponing the Inevitable?

The Vasai-Virar subregion consists of two major zones, the G zone (green zone) and the U zone (urbanizable zone). As mentioned earlier, the region was primarily meant to be the green belt for Mumbai. Even in 1995, the green zone covered 28,646.79 hectares while the urbanizable area covered only 9,343.21 hectares. As long as this ratio was maintained there was no imminent danger to the ecology of the region. The cause for concern presented itself in 1995 when the Special Planning Authority (SPA)[8], while taking up the task of formulating a sustainable land use plan, proposed a number of changes towards making more land "urbanizable". The proposal clearly identified that the region will be subject to residential, commercial and industrial development. The residential land use is to increase from 16.57 per cent of the total urbanizable area to 32.8 per cent (Table. 1).

Table 1: Existing & Proposed Land Use Changes for Urbanizable Area, VVSR

S. N.	Land Use for Urbanisable Zone	Existing Area (% of total)	Proposed Area (% of total)	Percentage Change
1	Residential	16.57	32.80	16.23
2	Commercial	0.80	6.16	5.36
3	Industrial	1.70	9.53	7.83
4	Social facilities	1.54	4.51	2.97
5	Public Utilities	0.13	2.72	2.59
6	Recreational Space	1.04	9.22	8.18
7	Roads & Railways	4.29	16.37	12.08
8	Forest	2.00	0	- 2.00
9	Water Bodies	2.50	13.01	10.51
10	Salt Pans	11.39	0.22	- 11.17
11	Agriculture	58.13	5.46	- 52.67
	Total	**100**	**100**	

(Total Urbanisable Area = 9353.21 h.a)

Source: Draft Development Plan for VVSR, CIDCO, 1995

Table.2. Existing Land Use & Proposed Changes for Green Zone, VVSR

	Land Use For Green Zone	Existing area (% of total area)	Proposed area (% of total area)	Percentage Change
1	Congested/huts	0.40	1.35	0.95
2	Roads	0.60	2.43	1.83
3	Railways	0.01	0.12	0.11
4	Water Bodies	0.43	3.50	3.07
5	Cattle Sheds	0.05	0.63	0.58
6	Forest	30.11	29.58	- 0.04
7	Agriculture	68.43	61.57	- 5.86
8	Garbage Dump	0	0.82	0.82
	Total	**100**	**100**	

(Total Area under Green Zone =28646.79 ha)
Source: Draft Development Plan for VVSR, CIDCO, 1995

The most irate feeling among the residents are evoked by the prospect of increased industrial activity in the area – 9.53 per cent land use has been proposed for industrial development, as against the existing 1.7 per cent. Overall, the proposal clearly indicated that 72.09 per cent of the U zone would be completely built up, while 9.22 per cent would be designated as open space for recreation and other activities. The area comprising water bodies, hills, salt pans and grazing lands is fast diminishing. Forest land is to disappear totally, according to the proposed plan.

In the green zone too, residential zones, garbage dumps, roads and railways are to take up about 5 per cent of the area (Table 2), whereas no construction should be taking place here. What is worrying is the fact that development is not a process that can be stopped all of a sudden; it gradually eats into adjacent lands that were once forests, hills, streams and marshes. Sustainable solutions are expected to curb growth within the rational framework of environmental concern. Where planning authorities falter, public participation may provide a clear direction to the goals of planning.

Access to Infrastructure and Environmental Quality

Alongside the issue of land use changes, it is also important to assess the availability of and access to basic infrastructure for the residential sector of the region. This will provide an insight into some of the related environmental problems that are likely to confront the residents.

Housing

The average annual growth rate of the subregion is 4.9 per cent and for the U zone, it is 5.8 per cent. The number of occupied houses rose from 28,600 units in the 1981 Census to 70,875 units according to the 1991 Census. The draft development plan for VVSR formulated by CIDCO revealed that in 1995, there were 83,808 housing units. The growing demand for housing is being met with by an equally rapid supply despite different kinds of regulations. The average occupation density is 4.69; and at this rate, the total requirement for new housing units by 2011 will be an estimated 5.56 lakhs.

Most of the units have come up recently, so the quality of construction and the conditions of the buildings are both satisfactory. Individual houses/bungalows are characteristically limited in number. The most commonly found tenement types include two-room units (65 per cent) and one room units (25 per cent); only 10 per cent of the units have more than two rooms. On an average, the housing units vary from 250 to 450 sq.ft, mostly suitable for lower-and middle-income groups. Tenement density (calculated using the gross residential area and total housing stock) is about 46 units per hectare and is more or less uniform. Despite the present trend of granting 10,000 to 20,000 permits for construction, there will still be a shortfall of housing units given the trends of population growth. Thus, the pressure on the limited carrying capacity will continue to increase. The region is indeed heading for disastrous consequences if authorities continue developing the area and builders persist with the building spree to meet the spiralling demand.

The CIDCO statistics also reveal that about *60 hectares of land is being cleared annually* for residential development. Thanks to the rapid rate of construction, the hills have been largely reduced to plains. With construction material piling up in the surrounding areas, the residents complain about the evident pollution. Sixty-seven per cent of the surveyed population feels that construction activity has led to environmental degradation. Construction materials are not removed from the sites. Eighty-nine per cent of the residents surveyed strongly feel that the developers must take responsibility for the removal of rubble, mud, concrete, ironware and cement mixes. All these are left untouched as the workers move on to a different construction site in an adjacent area.

Basic Amenities

The civic amenities that are basic to the well-being of any region includes water supply, drainage, sewerage and sanitation, solid waste disposal, roads, electricity etc. The haphazard development of Vasai-Virar has led to an alarming lack of these basic amenities.

Access to Water Supply: Of all the basic amenities, the most critical for VVSR is water supply. While the Maharashtra Water Supply & Sewerage Board has set the minimum requirement for urban areas at 110 – 180 lpcd, the available water supply for VVSR is a meagre 30 lpcd. Table. 3 shows the demand and supply condition of the region. In 1994 CIDCO acknowledged in its draft interim plan document that it did not have funds for the water supply schemes drawn on paper.

Table 3: Demand & Anticipated Supply of Water for VVSR.

All figures in mld of water

S N	Year	Total Water Demand	Drinking Water Required	Available Piped Water)	Available Well Water	Total Water Availabl e
1	1993	76	54	11	15 - 20	26 -31
2	1997	151	108.9	11	15 - 20	26 - 31
3	2001	226	162	11+130*	15 - 20	156 - 161
4	2011	380	270	11+130*	15 - 20	156 - 161

Source: Draft development Plan, CIDCO

* Possible yield from various sources:

a) Usgaon Scheme under execution: 20 mld

b) Surya Scheme proposed: 100 mld

c).Rajivali & Sativali Schemes: 10 mld

Water sources for most of the proposed schemes are situated at great distances from the main residential area and are likely to involve very high overhead operational costs. The supply from two main sources that presently exist – in the Pelhar Dam, which supplies 9.6 mld, and the Papadkhind Dam, with a capacity of 0.6 mld, is obviously insufficient. There are around 500 bore wells on the western side of the Western Railways. Residents opine that the eastern parts face a more critical water crisis. Nalasopara East has the most acute problem. Water is made available to the residents by tankers every alternate day, and the prices charged are unusually high. Residents even mentioned a tanker mafia that controls the supply and charges as high as Rs. 400-600 per month (for 15 days' supply). The water available is of poor quality; the hardness of water makes domestic use difficult. A very small section, comprising about 35 per cent of the total population, is able to avail of tap water- that too an irregular supply. A huge ninety-three per cent of the residents get tanker water every alternate day in Nalasopara and Virar; seven per cent of the residents get this supply every two days. Only eight per cent of the population jointly bear the cost of water with the housing societies; the remaining ninety-two per cent bear the full cost.

The indiscriminate withdrawing of water for construction work as well as for supplying to residents has been posing a continuous threat to the water balance at the macrolevel. The number of bore wells with sweet water that belonged to the farmers residing on the margins of the VVSR has also been dwindling. Tanker suppliers purchase this sweet water by paying a minimal amount to the owners of the wells. Approximately 5-7 million litres of water are pumped out of the groundwater reserve by tankers alone. In addition, 83.3 per cent of the farmers surveyed have also reported the seepage of saline water into their wells. The region away from the coastal zone is likely to be more seriously affected by the rapid depletion of groundwater in the very near future.

Sewerage and Solid Waste Disposal: At present there is no central sewerage system for the residential area. Individual owners have constructed septic tanks with soak pits. The soak pits are cleaned manually. The condition is likely to deteriorate rapidly as the population density increases and there is a proportional increase in the volume of waste generated. Besides, the manual cleaning of soak pits poses an imminent health hazard. It is evident from the survey that only a very small strip of land actually has the capacity to absorb the material collected in soak pits. The soak pits do not function well on marine clay and the effluents flow out into the low-lying areas. Small pools of effluents that accumulate in such areas pose a health hazard by supporting disease vectors and giving off an unpleasant pungent smell. A regulated mechanism needs to be evolved for water treatment and cleaning of the soak pits or else the situation will continue to be hazardous to the environment. The scientific solution for disposal lies in the construction of landfills, which cannot be considered at present due to the paucity of funds.

Uncontrolled solid waste disposal has been causing similar concerns among the resident population. A mere 20 per cent of the residents has access to public garbage bins and vats. Open space is used for disposal, with the waste dumps posing a threat to the environmental health of the population. Eighty-nine per cent of the residents feel that the disposal method is totally unsatisfactory. Uncontrolled and random disposal can have a dangerous impact on the region, which is largely a green zone. There has been an attempt to take corrective measures – the SPA has demarcated seven sites for landfills in the green zone itself. This can have negative repercussions in case the landfills are not scientifically managed. The most common problem such landfills pose is seepage from garbage dumps, which contaminates groundwater and pollutes the water table. It is well-known that the scientific management of landfills greatly reduces such risks, but these methods are cost intensive. Development authorities in most cases remain unable to bear the costs for varied reasons. Creating landfills in the green zone runs the risk of harming the local ecology, as in all probability they will not be scientifically managed. During the monsoons, landfills in Mumbai too are often found to be in a disastrous condition. Waterlogging and muddy surfaces discourage garbage trucks from driving into the fill, and the garbage is simply dumped on the approach roads near the entrance of the fill. This negates the justification for creating landfills for the management of waste disposal.

Road Conditions: The main transport route linking the region to the main city is the Western Railway line terminating at Virar, thereby connecting Naigaon, Vasai Road and Nalasopara. At the Vasai-Diva section, the Central Railway joins the Western but is mostly used for the movement of goods. National highway No. 8 (NH 8) connecting Mumbai to Ahmedabad passes through the east of VVSR. State highways SH-40 and SH-41 and the major district road MDR-39 connect NH 8. The condition of the intra regional roads is deplorable. Intraurban roads are few and inadequately maintained. Vehicles, pedestrians, cyclists all use the same roads, and footpaths are absent. There are few buses plying. A few housing societies have been developed several kilometers away from the station; these societies provide regular bus services to and from the housing complex and the station. There is no such facility for individual residents in the remaining subregion. Illegal structures have occupied land allotted for road construction. Ninety-three per cent of the residents surveyed indicated that while the area adjacent to the railway stations has no transport problems, interior areas are affected by the absence of roads and means of transport. People required to reach the railway station to avail of the suburban rail transit have to wait for anywhere between 1.5 to 3 hours for a bus to take them to the station. During the monsoons, the waterlogging in the congested areas makes the situation worse for pedestrians. It may be noted here that the SPA has made it very clear that it will take up the construction of only a limited number of peripheral roads, which will be more than 20m wide, for the entire region. All other internal roads will be the responsibility of the builders. The question is whether builders will be willing to invest in infrastructural work unless it is made mandatory.

Residents are of the opinion that the changing land uses are likely to cause severe damage to the ecology of this unique coastal area. Except for the air quality, all the other environmental parameters present a disturbing picture. This could become even more grim with land increasingly being put to industrial uses. Sixty-seven per cent of the residents surveyed suggest that environmental degradation is due to the recent construction activity. Some nearby villagers mentioned that before 1970, 45 per cent of the region was vegetated and 50 per cent consisted of water bodies, barren land, agricultural land, etc.

Conclusions

The problems discussed above are not unique to Vasai-Virar. Similar challenges face many other metropolitan regions that are experiencing unmitigated urban sprawl and are surrounded by ecologically sensitive areas like forests, wetlands, coastal zones, etc. Environmental concerns need to find their rightful place in the planning process. Human development; community welfare and social equity depend on flourishing ecosystems. Planning must face two basic realities – first, human uses of the environment will continue at a pace that doesn't match the carrying capacity; and second communities must thrive on the "interest" generated by the environment and not erode the "capital".

In this case, as also elsewhere, the primary concern is the Regulatory Mechanism. Only an effective regulatory mechanism can bring about sustainable development in these areas. The chief regulatory aspects are zoning, building codes, transfer of development rights and protection of critical areas.

The other important factor is the *scientific assessment of the vulnerability* of the region to degradation. This requires detailed studies to be conducted much before the area is finally selected for development. Environmental impact studies can indicate the carrying capacity of the region. At the same time, the availability of basic environmental services must precede the occupation of the area as poor supply of these services creates more stress by encouraging misuse.

In order to develop a set of scientific and sustainable planning objectives, the local development authorities must *develop a corporate approach* towards urban environment management. Since urban areas are intensive consumers of resources, the planning must be sensitive to conservation issues. As Andrew Blowers, Chairperson of TCPA, in the Report of the Working Group for Planning for Sustainable Development rightly states, "*Environmental planning is a comprehensive approach to environmental management, which has three basic features; it takes account of future uncertainty by a precautionary approach in the present; it reflects the integrated nature of environmental processes and policies; and it takes a strategic approach to decision-making* (Blowers, 1992:134)." All said and done, the success of urban planners depends on the commitment to sustainability and their ability to combine three important roles – that of mediator, strategic planner and regulator. The ability to operationalize sustainability within a complex network of actors, developers, agencies, and political and administrative contexts poses a challenge.

Notwithstanding the fact that the study was an attempt to highlight some of the critical issues for urban management, it must be acknowledged that this paper has been able to touch upon only a limited number of aspects. Each urban locale has its unique features and must be tackled in a specific manner, but the governing principles of sustainability must prevail.

NOTES

1. Vasai-Virar subregion or VVSR was specially delineated as a planning subregion by the development authority and assigned under the Special Planning Authority (refer to 8). Thus, the region is referred to as a subregion or VVSR in all documents, as also in the Draft Development Plan Document prepared for the region by CIDCO.

2. Green zone (G zone) is the area demarcated for preservation as a natural region where construction and/or land use change is prohibited. The Vasai-Virar subregion is divided into two zones, the green zone and the urbanizable zone (refer to 6).

3. Western Ghats is the term used to describe the stretch of uninterrupted hilly tract (elevations vary from 900 to 1,600 meters) that runs along the western coast of the Indian peninsula from the Vindhyan region to the southern tip of the peninsula.

4. Taluka is the term used to describe the administrative unit below the district level in most states of southern India. In northern states like Uttar Pradesh, it is known as *tehsil* while in West Bengal it is called block.

5. Urban Land (Ceiling and Regulation) Act (ULCRA) was adopted in 1976 by the Indian Parliament in pursuance of Clause (i) of Article 252 of the Indian Constitution. It was adopted in response to:

a) a demand for imposing a ceiling on urban property, especially after such imposition was made on agricultural land by the state governments.
b) a need for orderly development of urban areas as increased urbanization and population growth took place.
c) the necessity for a measure to exercise social control over the scarce resource of urban land.

6. Urbanizable zone (U zone) is the demarcated area of the Vasai-Virar sub-region that can be converted for urban uses such as residential, commercial, infrastructure development, etc. This is different from the G zone as construction is permitted and developmental activities may be carried out in this zone.

7. Gram Panchayat is the lowest administrative unit in the three-tier local self-government system in rural areas. ('Gram' means village, panchayat implies elected administrative body)

8. Special Planning Authority (SPA) was created by the Mumbai Metropolitan Regional Development Authority (MMRDA) with the special responsibility to look after the planning and development of areas earmarked for them from time to time.

REFERENCES

Bandyopadhyay S 1998, *Urban Space Use in the Residential Sector, Case Study: Vasai-Virar Sub-Region, Maharashtra,* MPhil dissertation submitted to Indian Institute of Technology, Mumbai (unpublished).

Blowers A 1992, "Planning for a Sustainable Society", *Streetwise,* Spring, No.10.

Commission of the European Communities 1990, *Green Paper on the Urban Environment,* Commission of the European Communities, Brussels.

City and Industrial Development Corporation (CIDCO) 1995, *Draft Development Plan for Vasai-Virar Sub-Region,* CIDCO, Mumbai.

Fernandes A 1989, "Vasai Regional Plan", *Times of India,* 23 September 1989, Bombay.

Grant J P and Manuel & Darrell Joudrey 1996, "A Framework for Planning Sustainable Residential Landscapes", *American Planning Association Journal,* Summer, vol.62, no.3.

Sharma R N 1991, Land-Grab, Bombay Style: "Urban Development in Vasai-Virar Hinterland of Bombay", *Economic & Political Weekly,* February.

UNCHS (Habitat) 1986, *Global Report on Human Settlements,* United Nations, New York.

Yoo Jae Hun 1991, The Conflict Between Urban Land Development & Green Area Preservation: Case of Seoul, Paper presented at the International Conference of Dynamic Transformation of Societies.

18

Ushering in Clean and Green Urban Centres

The Experience of Hyderabad

N Ashok Kumar

Introduction

Unsustainable growth is plaguing urban India. Population explosion has outstripped the institutional capacity of the urban local bodies. Relentless increase in the urban population is heavily taxing the existing basic civic amenities. Judicious environmental management has thus become more relevant than ever before. Handling of solid waste, for example, is a major challenge. Urban local bodies (ULBs) are grappling with the declining quality of the services provided by them. With a view to addressing these problems, municipal bodies have been trying to work in partnership with citizens through a participatory process. The Andhra Pradesh government has initiated a path-breaking strategy to make Hyderabad and other cities and towns in the state green and clean.

Hyderabad city has been facing a multifaceted crisis. Many a public place was deprived of a clean environment. The inhabitants of the city often felt weighed down by the gravity of the civic problems that plagued Hyderabad. It looked so pathetic, with unpaved streets and unclean surroundings. The shoddy state of roads and sanitation was also true of other services. The inescapable conclusion was that the city has grown beyond its capacity, in terms of provision of civic services to its citizens.

The focus of this paper is Clean and Green – a populist environmental improvement programme, which the Andhra Pradesh (AP) government has been implementing since 1998 both in Hyderabad and other cities and towns of the state. This programme has become a blueprint for planning smart cities and towns in AP. Clean and Green has successfully demonstrated that economic development and environmental protection ought to go hand in hand to make the cities more sustainable.

Green City

Once Hyderabad was known for its gardens, orchards and water bodies. There used to be more than 500 water bodies in and around the city. Today there are only 150. Over the years, Hyderabad's physical environment fell into a state of steady decline. It still has five major parks and 240 small parks. A poor infrastructural network and severe resource depletion were negatively contributing to the local economy of the city and undermining its ecological sustainability. New investments in the city were not giving adequate returns.

With a view to enhancing productivity and life in harmony with the environment, the state government launched two marathon populist programmes, namely, 'Janmabhoomi' (birth land) on 1 January 1997 and Clean and Green in August 1998. The latter sprouted as an offshoot for the former, with the shared vision of extending the green cover of urban settlements.

Till date, the state has gone through 15 rounds of the Janmabhoomi programme. The latest was during 1-7 June 2001. During these fifteen rounds, various kinds of socio-economic and environmental improvement activities were taken up. These include rainwater harvesting, sanitation works, primary education, environment conservation, physically handicapped persons' welfare, primary health and family welfare, responsive governance, women's empowerment and farmers' welfare. A people-centered participatory process is the base of this programme.

With regard to funding of the programme people's contribution constitutes 30 per cent. In the case of urban slums and downtrodden communities the share is halved to 15 per cent. The balance amount is provided by the government. Those with hands-on experience have noted certain shortcomings in the implementation process. It appear these populist programme have been implemented without any action plan. Opposition parties argue that they were not given the opportunity to participate in these programmes. This was markedly witnessed during the 14th round of the Janmabhoomi programme, which had as its focus women empowerment. On account of this charge it is learnt that both officials and politicians faced the ire of the people during the last two rounds of the programme. In spite of weaknesses, the Janmabhoomi programme has been continuously implemented without a break since it was introduced.

The concept of green Hyderabad has emerged as a model to enhance the physical environment of the city. The Andhra Pradesh government conducted a mass tree plantation programme in Hyderabad in August 1998 and from 10 September 1998 in other cities and towns. There is no financial provision for this programme. As on 20 July 2001, 33 rounds of this programme had been conducted throughout the state. During these rounds, much importance was given to increasing the green cover in neighbourhoods and creating awareness about the clean and green concept among the people. As much as Rs.3027.9 million worth of work have been undertaken in the area of sanitation during the programme implementation. Users contributed in the form of cash, labour and kind.

Implementing Agencies

Municipal Corporation of Hyderabad (MCH), Hyderabad Urban Development Authority (HUDA), Hyderabad Metropolitan Water Supply and Sewerage Board (HMWS&SB) are the major implementing agencies of this programme. Each department was entrusted with specific tasks. HUDA was made responsible for plantation of seedlings and their maintenance; creation of water bodies and water fountains; development and maintenance of parks; construction of fly-overs; and landscaping of gardens beneath the fly-overs. Similarly tasks such as keeping the city clean; creating awareness among citizens about the programme; social mobilization; organizing coordination committee meetings; and plantation and development of parks were assigned to the MCH. Landscape gardening, orchards cultivation near water bodies, creation of new water bodies and rain watershed works were vested with the HWS&SB. Development of social/urban forestry was the job of the urban forestry wing of the HUDA.

The Hyderabad Urban Development Authority was constituted in the year 1975. In 1986, the urban forestry unit was created within HUDA. Though, it had been involving in tree plantation for a long time, its initiatives were piecemeal and scattered. However, under the 'green' programme HUDA has formulated a policy to make the city lush green, improve its natural climatic conditions, halt the increasing pollution levels and maintain the hydrological cycle – in short, a policy informed by a sense of environmental responsibility. Towards this end, it has commissioned a long-term project known as "Green Belt." Under this project, HUDA plans to convert all the unproductive lands into greenery and safeguard the natural resources underpinning the provision of daily wage employment to the urban poor. This project has been implemented in two phases. The Royal Netherlands government has funded this project.

Phase I

Phase I of the Green Belt project was launched during 1993-94, with an estimated cost of Rs.996 lakhs. Its duration was five years (1994-99). As per the MoU, the Netherlands government bears 75 per cent of the total cost, with the state government contributing the rest. The estimated area covered by the project was 5,000 ha, which included mostly the foreshores and catchment areas of all major water tanks, residential and industrial, wasteland, institutional open areas and brown fields. Works such as conversion of 1,400 ha of private wasteland (belonging to farmers) into agro-based farm forestry and creation of daily wage employment for the urban poor to the tune of 20 lakh man-days were taken up. The other major work undertaken during this phase was the conversion of unproductive lands into greenery. In addition, works like roadside avenue plantations, consisting mostly of silvi-pasture species, agro-forestry plantations, water supply plantation, etc., were also taken up. Silvi-pasture species have tremendous potential because they provide welcome greenery as well as yield good returns for the growers.

Impact

Phase I had a runaway impact on the living conditions of the beneficiaries. Ninety per cent of the target was achieved during this phase. In its evaluation report, the Dutch government's Review Mission stated, "The present scheme has proved to be highly successful. The survival rate has been around 95 per cent." The Indian government gave HUDA the Tree Protection award for the excellent work done in reconnecting the city-dwellers to nature. Appreciating the results, the Netherlands government provided additional finances to the tune of Rs.10 million for extending the duration of Phase I by a couple of years. Additionally, the Dutch government agreed to provide financial assistance for Phase II of green belt project.

Phase II (1999-2004)

Phase II of the project was launched in 1999. Various works like conservation of lakes, rehabilitation of those affected by the lake beautification work, tree and roadside avenue plantation, beautification of water bodies/water tanks, developing parks and landscape gardens are the major components of this phase. The total cost of Phase II is estimated to be Rs.2210 million. The duration of this project is expected to be 5-10 years. In the initial stages, HUDA has proposed to undertake reconstruction works like the removal of silt and mud from water tanks, development of sewage farms for artificial purification of sewage, construction of bunds/causeways to preserve water, development of parks around the water tanks, cleaning of drains in about 25 tanks out of the total existing 110 tanks within its jurisdiction. These projects are under implementation.

Citizen's involvement in project implementation was given paramount importance. To set the project into motion, urban forestry officials held on-site ward committee meetings involving community-based organizations and residential welfare associations. During these meetings, officials shared their ideas with the people about the project implementation process. It was quite a novel experience. In the beginning the people's response was lukewarm. This led the urban forestry officials to conduct acceptance workshops and publicity camps in the neighborhoods, with a view to motivating the people to become actively involved in the programme implementation. This exercise has markedly enhanced the people's involvement – particularly of the urban poor, women and school children. Women's groups have taken up voluntary work, such as raising nurseries and cultivating fodder crops on barren lands/open spaces. This has greatly supplemented their family incomes. In all these activities, HUDA played only a facilitative role. As part of this project, about 10,000 beneficiaries were supplied with seedlings and tree guards free of cost. Additionally, financial support to undertake works like digging pits for planting seedlings and watering thereafter was also extended to the beneficiaries. This novel experiment has reaffirmed the fact that, acceptance is a prerequisite for successful social mobilization.

Subsequently, HUDA launched another environmental project, namely, Green Hyderabad. The major aim of this project is to halt the increasing pollution levels. It has a budget of Rs.1.6 million during the inception phase.

Other Components of the Green Package

As part of phase II, the urban forestry wing has taken up other jobs like block plantations and highway strip plantations; development of parks, traffic islands, lake banks, central nurseries, seedling producing centres; developing picturesque sprawling gardens in public institutions like hospitals and office buildings, including the state secretariat building; developing a national park, greenhouses and landscape gardens beneath fly-overs. All these activities had given facelift to the physical environment of the city. Of these, the landscape gardens beneath the fly-overs are especially pleasing to the eye.

Subsequently, HUDA has incorporated certain changes in the project implementation process. Unlike in the past, today, stakeholders have to bear 50 per cent of the cost of the tree guard and the manual labour charges for transporting, transplanting and watering seedlings.

Pursuant to the very encouraging results from its initial experiments, HUDA embarked in August 1997 on another unique experiment, called One Day One Lakh Plantation Programme in about 17 places in the city. COVA (Conference of Voluntary Agencies) took the lead in a big way. It supported the programme financially and propagated this novel experiment among schoolchildren and women.

Box. 1: **Participation of Women's Groups – Success Story 1**

Saktimahila Sangam is a women's forum with one hundred groups. These women's groups were permitted to use 5 acres of barren tract of government land for the cultivation of a special genus of silvi-pasture species, known as *Stylo santhus Hamata*, in a village called Mamidipally, 5 miles away from the city. This species is more nutritive for cattle and improves the yield of milk. Further, the crop yield is plentiful. Women cut the leaves of this plant at regular intervals, which they sell in the market. The seeds of this plant fetch a lump sum income to the growers. Cultivation of this species also helps in the preservation of the ecological balance of the city and keeps the day temperature during summers under control.

Apart from this species of plant, the women's groups also grow the *Cytisus* or *Genista* genus of shrub and other medicinal plants in the barren lands surrounding their living environment. They also grow broom shrubs and use the broomsticks to make sweeping implements. All these activities bring in additional income for the families of the women groups.

Additionally, HUDA has adopted 17 slums. It had provided training to slum-dwellers in homestead nurseries, development of small water bodies, avenue plantation, cultivation of seedlings of flowering, ornamental, vegetable and fruit-bearing plants and slivi-pasture. The urban forestry wing itself is the prime buyer of the seedlings raised by the slum-dwellers on a 'buy-back' principle, thus enabling the urban poor to keep the home fires burning. It is learnt that, each household is generating Rs.400-500 per month. Further, local women's groups are cultivating slivi-pasture species (cattle fodder crop) in the unused barren tract of lands/vacant government lands through cooperatives. This is a unique experiment.

The state government has asked HUDA to raise the city's green cover from the present target of 15 per cent of its 190 sq. km area to 30 per cent over the next three years. The government plans to enact a law to popularize the 'green building'. It has asked builders to conform to global environmental standards for the construction industry.

Clean City

Municipal Corporation of Hyderabad (MCH) has launched the Our City-Clean City project. Capacity building, increasing public awareness and people's involvement creating a healthy and clean environment are the major components of this project. Awareness generation among the citizens is carried out through mass education media. MCH has put up hoardings at select places emblazoned with slogans like Grow More Trees and Make the City Green; Clean and Green For Everyone and by Everyone; to encourage people to participate in the clean and green programme. Further, the municipal corporation has dovetailed this project with the Janmabhoomi programme for more effective implementation.

Clean Slum was the other initiative taken up by MCH. Women's groups in the slums were identified and trained to function as mini municipalities for collecting contributions from households for maintenance of sanitation in the neighbourhoods. Each group was supplied with a tricycle for the collection and disposal of garbage. The corporation bears 50 per cent of the total salary expenses on account of the conservancy staff engaged by the slum-dwellers, including the salary of the tricycle puller.

Impact of our City Clean City

The changes the project has brought about are there for all to see. Solid waste collection which earlier was not more than 20 per cent has shot up to 85-90 per cent. MCH is now regularly black topping and widening roads, removing illegal constructions, lifting garbage, constructing public toilets and keeping them clean, sweeping the streets, etc. Most people agree that Hyderabad's physical environment has improved considerably.

Public awareness about civic issues has increased. The television programme *Face to Face* recently introduced by the Andhra Pradesh government has played an important role. A private TV channel telecasts this programme every Wednesday, allowing citizens to interact with the concerned officers on various civic problems. Like *Face to Face* TWINS (Twin cities integrated network services) is also a programme in meet-the-people genre. During this live TV programme, citizens share their day-to-day problems with the concerned government personnel. People have access to eighteen different departments.

However, the city is not totally free from environmental problems despite the municipal corporation and the state

Box. 2: **Participation of Women's Groups – Success Story II**

Women's groups have been involved in the Clean and Green programme in Kukatpally municipality on the periphery of Hyderabad Municipal Corporation. Women's groups formed under DWCUA are successfully undertaking sanitation works having developed access to self-employment opportunities under this programme. Since this was the first time the municipality was experimenting with such an initiative, the groups were initially given training in solid waste management. The women's groups were involved in more than one area of operation, which included building group leadership, becoming contractors and undertaking civic works, conducting the credit and thrift activities, mobilizing contributions from residents for various community works, etc. The municipality has contracted out basic sanitation works like cleaning roads, clearing bushes and lifting solid waste to the women's groups. By involving themselves in such initiatives, women have defied the conventional contractors and local leaders' whips and threats against their leading role in the development of their neighbourhoods. Further, these women have helped to improve the economic conditions of their families.

In Uppal, another municipality on the periphery of the city corporation too women's groups were involved to enthuse their fellow citizens and motivate them to contribute small amounts for community works.

government having taken up the initiatives discussed above. Hyderabad still has to contend with littering and spitting in public places; inadequacy of public toilets for men and absence of public toilets for women; uprooting of trees to facilitate road widening; pollution and traffic jams. Citizens of the city too are responsible for this state of affairs. At time, MCH has been compelled to discontinue its innovative initiatives because of a few intransigent people's activities. Genuine transformation of people's attitudes with regard to keeping the surroundings clean is a must.

Public – Private Partnership

Public and private agencies have also taken initiatives to improve the green cover in the city. The railways have been battling the problem of encroachment on its vacant lands in Secunderabad division by land sharks. To check this menace, the railways has planned to transform its vacant lands into beautiful landscape gardens, parks and lawns. It is a path-breaking experiment. There is no exclusive financial allocation for this project. The only materials used for these works are scrap iron and used wood blocks. Employees will volunteer physical labour. These projects are to be entrusted to Indian Oil Corporation for regular maintenance.

The state tourism department has involved private firms in different eco-tourism projects. During the first week of August 2001, the AP forest and tourism departments established the Botanical Gardens (phase-I) in an area of 27 acres, out of the total project area of 120 acres, in Jubilee Hills. Major traffic islands in the city are being operated and maintained by major private firms.

Implementation and Achievements

A high-level coordination committee was constituted at the city level by the government to monitor the implementation of the Clean and Green programme. The Commissioner of MCH is its coordinator while the chairpersons of HUDA, HMWS&SB and AP Transco are its members. The committee meets to reviews the programme implementation once a week.

Though there has been some loss in green cover, Hyderabad (once a city with indifferent civic sense) has won HUDCO's "Clean City" Award for three consecutive years. The state stood first in improvement of green cover, with its share occupying 965 of the 3,000 sq km in the country.

It is not out of place to mention that the schemes for greening the city have not only improved its physical environment but also provided casual and/or self-employment opportunities to the urban poor and women's groups. Many among the poor have become small entrepreneurs. They manufacture lawn mowers, lawn sprinklers, lawn sand and tree guards, besides raising seedlings of various flower, vegetable and fruit plants. Many of them are engaged as casual labourers in construction activities, like building tree guards, developing landscape gardens, creating lawns on traffic islands, and as watermen.

The clean and green experiment of Hyderabad has successfully demonstrated that local solutions are needed for safeguarding natural resources, on the one hand, and for strengthening the economy and enhancing the quality of life of the people on the other.

PART III

Appendices

Appendix 1

Working Group Reports

Group - I

***Theme:* Access to Environmental Services and Environmental Implications.**

Four determinates of access to environmental services emerged from the group discussion. First, the spatial characteristics, that is, the physical space from where the services are to be accessed were taken up. Here the focus is primarily on size, density, location and other spatial characteristics. Second, the population characteristics of the community for whom the services are meant were discussed. These include the income range of the concerned population, its cultural characteristics, the extent of migration and community issues. The third determinate of access to environmental services discussed by the group was technology. Of particular importance for the working group was the fourth determinate – the political dimension. Here, questions like a democratic versus non-democratic political system and a decentralized versus centralized political structure become important.

Among other aspects that came up for discussion related to people's participation, market versus the state, management and enforcement, and laws and legislation and jurisdictional questions. These issues were discussed in the context of a whole range of urban entities – a city, a region, municipalities and collection of cities. Here it was pointed out that the absence of government participation leads to undue dependence on the market. This has three serious environmental implications. First, overexploitation and depletion of natural resources like land and water; second, encroachment of public land; and third, high levels of pollution, which if not irreversible, demand prohibitive investments for mitigating the damage.

On policy-related issues, the discussions revolved around five aspects: (*a*) planning and control; (*b*) budgetary measures and issues of subsidy; (*c*) public action and people's participation; (*d*) appropriate linkages between various complementary sectors of the economy; (*e*) scope for improving institutional capacity.

The group advocated adopting a holistic approach on the question of planning. Some participants mentioned the Brazilian model, which reflects a strong city government and a city parliament as an appropriate example. All agreed that it was necessary to have development councils from the local to the national level. The group also mentioned that a prior land-use planning exercise must be undertaken in order to prevent unplanned growth. It suggested that environmental impact assessments should be carried out for both rural and urban areas so as to bring transparency to the planning process. In the context of linkages among various complementary sectors, capacity building of the institutions involved in the provision of basic services was emphasized.

Group II

Theme: Institutional, Fiscal and Legal Issues for Environmantal Management with Special Reference to Urban Local Government

At the outset, the group decided that along with institutional, fiscal and legal questions, constitutional issues also needed to be addressed. It was pointed out that though some beginning had been made in India in the area of constitutional reforms for empowering municipal governments, the constitutional provisions in other countries were not adequate and required closer examination. For example, in Bangladesh, "local government" is mentioned in the constitution but not "urban local government." In Sri Lanka, a three-tier system was created, as in India – that is, national, provincial and local governments. And in both countries, local government is a subject on which provincial governments legislate.

Highlighting the Indian example of the 74th Constitutional Amendment Act (CAA) the group argued that powers cannot be devolved in the true sense to urban local bodies (ULBs) mere by constitutional amendments. The various states have yet to make suitable changes in the legal framework, following the passage of the 74th CAA, with respect to fiscal matters. The entire context of municipal legislation has changed. Traditionally, municipalities were considered to be the launching ground for young politicians. Today, ULBs are supposed to be manned by professional managers. It was further pointed out that the 74th CAA is basically inspired by the 73rd Constitutional Amendment and hence the real concerns of urban areas have been missed, that the constitutional provisions have not appropriately devolved at the state government level. For example, there is no list of activities (items) that is specifically mentioned within the gamut of operations of local self-governments (corresponding to the central and state lists).

On legal issues, the group focused on the fact that there are a multiplicity of overlapping planning and municipal laws on environment protection and management. These laws have to be examined in detail so that an institutional structure for proper enforcement can be put in place.

Regarding institutional questions, the following points emerged from the discussion:

- Institutional tie-ups among all stakeholders are necessary so that measures to mitigate environmental damage can be initiated.
- Community participation is important; community support must be garnered in order to ensure ownership of the programme.
- Application of the strategic management concept in the area of environment by cities in Sri Lanka and Thailand can be adopted by other countries as well. This strategy is people-oriented and supports an institutional framework based on community participation and public-private partnership.
- There should be appropriate mechanisms and guidelines for implementing government programmes to enable them to achieve their objectives. NGOs can play an important role in sensitizing slums-dwellers to become more participatory in the process of development.

On the fiscal dimension, the group held intense discussions on municipal resource mobilization, cost recovery, subsidies, credibility of donor-driven programmes, etc. The following points emerged from the discussions on the several country-experiences cited by the participants.

- To fulfill the cost recovery agenda, capacity building of municipalities is imperative. The municipalities must tap the capital market to be able to provide urban services that are efficient and reliable.
- Suggested financial reforms of municipalities included better resource generation from existing sources, cost recovery and a transparent accounting system.
- Though international donor-driven models in urban environment management programmes are not the only effective models for improving the urban environment, external donor agencies contributed to a large extent towards the process of institutional capacity building.
- Given the heterogeneity of the urban sector, it would be worthwhile to make a distinction between categories of users requiring protection and those who have the purchasing power to access services from private markets.
- Empirical evidence suggests that municipalities in India have the potential to develop into self-sufficient institutions. Internal resource mobilization through the fullest exploitation of available means is the most important measure in this direction.
- Property tax, which is a major source of municipal income, is often not collected by municipalities in an efficient manner.
- While subsidies need to be provided for the poor, the process must be transparent.

Group III

Theme: Community Action and NGO Involvement in Urban Environment Management

The key questions that emerged from the discussions of this group were:

- Who sets priorities for a city and how are they decided?
- What types of participation need to be promoted?
- How can trusting, productive partnership be fostered?

The group arrived at consensus that definitions of terms such as civil society, CBOs, NGOs, private sector, environment, participation, etc., are important though definitions are contextual. It was decided that this forum would avoid defining terms.

The group identified the following prerequisites for constructive participation:

- Clear institutional structure for participation at all levels
- Definition of the processes of participation
- Changes in the behavioural culture to develop equal partnerships

The need for a significant change in mind-set at the municipal level, both among the bureaucracy and political leaders, was pointed out. This change should usher in a new culture where the citizen is seen as a partner in local governance – a partner who has rights as well as responsibilities. Examples in some Asian cities show that there is potential for bringing about such change.

The new culture will require a strong focus on communication and transparency in governance. Institutional mechanisms must be established for the exchange of views. Face to face meetings and other accessible mechanisms for communication need to be developed between senior levels of local government and citizens from all walks of life. Citizens must be able to present their demands and the government should be in a position to promptly respond. Such mutual exchange will not only allow problems to be aired but also encourage attempts to jointly work out solutions.

The group emphasized that every attempt should be made to develop a pluralistic political culture in municipal governance. Involving opposition politicians in initiatives, establishing cross-party committees, sharing challenges and successes among parties will encourage non-confrontational decision-making and promote the sustainability of effective measures – beyond the lifetime of one government.

Government decision-making should be taken as close to the people as possible. Local governments must decentralize planning, resource allocation, and decision-making within the city. Ward-level planning should be promoted; for example, the formulation of local environmental action plans, including the specification of resource requirements.

It was pointed out that there is a huge inequality in access to information and knowledge among different sections of the citizenry. There needs to be widespread awareness at the community level, particularly among poorer citizens.

Communities, even of the poorest, have resources, which should be acknowledged and used creatively in partnerships. The local knowledge of people living in cities is valuable. This should be utilized in local planning and decision making. Communities, at all levels must be aware of their rights as well as their responsibilities. Good will, cooperation and trust between the government and citizens can be generated through evidence of action and change.

The private sector should be asked to shoulder its share of responsibilities towards cities. Options could include sponsorship, managing parks and gardens, cost sharing for infrastructure, etc. Win-win options should be sought. Any privatization of services should not infringe on the democratic rights of citizens.

The group also drew attention to the role of NGOs and CBOs. They should be involved in developing innovative economic mechanisms to cover costs of service provision, both capital and recurrent. Every attempt should be made to build constructive relationship between NGOs with different constituencies, skill and priorities. Active communication within the NGO community is fundamental to creative partnerships between NGOs.

Stress was also laid on the need to monitor the impact of initiatives. All too often, government and NGOs tend to quantify the outputs, in terms of activities, coverage, targets, etc. Interventions can only be deemed to have had a sustainable impact if they lead to positive change after the input has ceased.

Appendix 2

Panelists' Responses

Prof. Christopher Benninger, renowned architect - planner and environmentalist, pointed out that in a democratic country like India, it is not fair to undermine the role of politicians. They are the ones who work with the community and in a way shape the voices and decisions of its members. Thus, politicians' participation too needs to be highlighted. Prof. Benninger talked about how politicians are trained and their working relationship with city managers. He noted that the profession of city manager is yet to be established in a real sense; neither were there institutions to provide them with professional training. He added that though a lot has been said about decentralisation of political power and devolution of finances, decentralisation of capability is rarely mentioned – an issue that is important. Prof. Benninger expressed his apprehensions about community participation. He argued that the urban poor should not be expected to invest time or money to get basic services as they are already quite hard pressed for survival. On the role of NGOs, his comment was that they tend to work in pockets in their own constituencies and are almost like parallel governments. While the participatory and decentralised approaches to development identify NGOs as major actors, the degree of transparency that informs their working methodology is debatable.

Mr. Prabh Das, Chief Executive Officer of Kolkatya Metropolitan Development Authority (KMDA) restricted his comments to West Bengal, especially Kolkata. Decentralization, he pointed out, was practised in West Bengal much before the implementation of the 74th Constitutional Amendment Act. He talked about the Mayor-in-Council system that is practised in Kolkata Municipal Corporation. He responded to the discussion that had been generated by one of the presentations on the relevance of KMDA in the present context, where powers are being devolved to urban local bodies. Mr. Das reiterated that KMDA even today has a significant role to play. There is no overlapping of roles between the KMDA and ULBs. Both these institutions have worked hand-in-hand for several years and there was never any hierarchy between them. It is simply that the intervention level of these institutions is different. ULBs cannot be expected to plan at the regional level – like designing an expressway, a railway network, or a major water supply network. They are better networked at the local level and perform better at this level. He cited the example of CUSP (Calcutta Urban Sciences for the Poor) programme, which KMDA is implementing in partnership with ULBs. He pointed out that although powers are devolved to ULBs, it is impossible for them to carry out the delegated tasks until and unless there is an effort to build their capacity. He argued that KMDA has technical capacity, which ULBs lack. Instead of replicating this capacity, ULBs should hire these professionals from KMDA. Mr Das shared with the group the initiatives taken by KMDA to help municipalities to computerize their operations. A pilot project is being run in 39 municipalities in West Bengal. On the question of subsidies, he pointed out that except for metro rail, which received subsidies 86 per cent of the transport in Kolkata is in private hands and thus is not subsidized. But, unlike transportation, water is a heavily subsidized sector in this city. In recent times, there has been talk about levying user charges on water supply.

Mr. Helmut Dedy, Executive Manager, German Federation of Cities and Communities responded to the issue of right to access to urban services, which was raised by the first group, and drew parallels with the situation in Germany. In his country, debates centre around the prices charged for public services like solid waste management and water supply. On the institutional question, he mentioned two aspects. First the different forms of self-government are practised in Germany and India. In Germany, the municipalities are guaranteed the right to manage their own affairs for which they are fully responsible. The terms of mayors are assured for at least four to five years, although this is currently being debated. The second aspect he drew attention to was financial constraints, which had forced municipalities to carry out tax reforms. Regarding community participation, Dedy mentioned that NGOs are perceived differently in terms of their role, in European countries and in India or Asia.

Open House Discussion

At the outset, **Mr. K C Sivaramakrishnan**, former Secretary, Ministry of Urban Development, Government of India, who chaired the session, clarified the role of district planning committees and context in which they were set up. Provisions for district planning committees were made in the 73rd and 74th Constitution Amendment bills, passed during the prime ministerial tenure of Rajiv Gandhi. There was a great deal of opposition to this from the Planning Commission. The Commission is created by an executive order, and it was feared that if the Constitution confers planning powers on any other entity, the authority of the Planning Commission would be undermined. Further, the role of the state planning machinery, which is also a part of the executive, would be diluted. This is why the terms "draft development plan" are used legally – implying that somebody else will take the final decision. He also pointed out that "development plan" can have several interpretations, and it would not have been possible to define the word "development" sufficiently within the Constitution.

Mr. Ashok Kumar commented on the presentation made by Group II. He said that instead of concentrating on the internal mobilization of resources, municipalities should first focus on the maintenance of assets. If this can be done efficiently, more revenue can be generated. Regarding subsidies, he observed that their withdrawal affects middle class the most. Citing his experience of Hyderabad city he pointed out that the rich often get away by not paying taxes. On the top-down approach that most ULBs practise, Kumar suggested that the initiative has to come from above for bringing about changes in the system.

Prof. Amitabh Kundu said that it is important to reflect on what enables capitalism to succeed in some countries and which factors impede globalization. Why are some countries adversely affected by globalization? He pointed to the legal system as one of the decisive factors. In a survey conducted in 10 to 12 less developed countries, it is found that countries with successful development policies are those that have very formal legal structures, which enable them to grow. Many developing countries are moving towards formalizing their systems. In this context, he shared the findings of a study on tenural rights by Geoffrey Payne. This study concluded that more than formalizing the systems in developing countries, the urban poor need to be given political and administrative space, which is being squeezed by economic growth. In other words, given the inequality that exists in cities, formalization hits the interests of the poor. By increasing the political, social and administrative space, formalization can work for the poor.

Ms.Alison Barrett initially responded to Christopher Benninger's comments on people's participation, where he argued that it is illogical to expect the poor to invest time and share the cost of providing urban services. Benninger had also said that for quite some time, people's participation was being talked about without much success; so may be, this calls for a change in direction. Ms. Barrett argued that although people's participation may not have achieved all that it hoped to, this was the only way to address urban issues. She said that these issues need to be raised more strongly; and attempts should be made to work towards win-win scenarios, which can show the way forward. There have been positive actions on the ground, including some city-wide dialogues and partnerships. These need to be highlighted as more successful examples come up each year. Ms Barrett pointed out that modern communication and information technologies, like GIS, web sites and television, could be used to document the success stories. It was not that no progress had been made; sometimes it is not sufficiently talked about.

Reacting to Prof. Amitabh Kundu's suggestion on formalization, she said it should become to a greater degree an expected part of governance. She also emphasized the need for more accountability on the part of politicians. Alison Barrett ended her comments by underlining the importance of the role of the community. Even small victories bring about a change in their attitude and culture and boost their confidence. The community can make a positive contribution in all aspects – be it planning or implementation, maintenance of finances.

Mr. Sivaramakrishnan said that when it comes to providing subsidies for protecting the environment, an accountant's perspective does not help. In Delhi, according to the 1999 statistics of the Ministry of Surface Transport, 680 motorized vehicles were being registered every day; the number now is probably 900 per day. In 1961, compared to Delhi's 16 per cent of road space, Kolkata had only 6 per cent. Who has taken over the city's roads? It is private automobiles. Are they being charged the right price for using the road space? What about the taxes on buses? Traffic and transportation, Sivaramakrishnan said, have to be considered together. It is not just a matter of accounting for bus fares. In cities across the world, certainly in developing countries, the proliferation of private transport is a very serious environmental problem. He did not believe that cities in India had either the competence or the guts to deal with this growing menace. At which sectors are investments, including private sector investments, targeted? Who gets the benefit? What is the real cost to the public exchequer? These are important questions that need to be addressed in the context of subsidies.

Mr.G M Ghorpade talked about the paradigm shift in planning. In the master plan paradigm, even 20 years after the formulation of a master plan, modifications can be made or it can be discarded. He cited the example of the Vasai-Virar regional plan, where during 13 years four drafts have already been made. Without political pressure, these plans are merely planning exercises. He argued that the planning process has to have technical inputs with a lot of managerial skills. Alternatively, the CDS approach presented by Perera is more appropriate. Where, in contrast to the typical development plan or master plan approach, all the actors – agencies, interest groups, opposing groups, pressure groups – are involved and a consensus is arrived at.

Mr. Ramaswamy's comments pertained to community action and NGOs. He said that there is a multiplicity of roles that NGOs can play. However, they need to coordinate their initiatives and act in concert. Despite years of social activism, in no major city of India, have NGOs been able to evolve a common perspective and be part of a single platform. The second issue he highlighted was the sharp divide between people who can access the urban infrastructure and those who cannot. He said that this gulf needs to be bridged first, with the initiative coming from those who are in a more advantageous position.

Mr. D.M. Sukthankar pointed out that much of the discussion during the conference had focused on the need for accountability and transparency in the working of the system. There is a debate about what instrumentalities are available and how they can be structured to ensure that in practice there is more openness, more access to information. Certain legal provisions, even if they are not perfect instruments, are being tried. He referred to the subsidy versus full-cost recovery debate and pointed out that in the Maharashtra municipal legislation, a provision has been made to deal with this particular aspect. The municipal commissioner, while presenting the budget, has to circulate a report to the house and the standing committee on the quantum of subsidies extended to various civic services. Take for example, water supply and sewerage. He has to indicate the costs of operation and maintenance, depreciation and debt servicing. This is the expenditure side. On the revenue side, information has to be shared on the actual receipts realized during cost recovery – whether by way of taxes, which is a percentage of the rateable, or directly by means of metering and on the basis of the quantum of service consumed and the service charge levied. The gap between expenditure and collections that is being subsidized from general revenues also has to be laid before the house. The municipal commissioner is obliged to tell the house who are the beneficiaries of this gap; in other words, who is paying what, and in what manner are these gaps being filled. This is the instrumentality that is provided in the legislation so that information becomes accessible to citizens at large. There can then be a public debate on the ways and means of bridging the gap and so on. Mr. Sukthankar also referred to the environmental status report, which again is required to be laid before the house. The process of coming out with a comprehensive report is an on going one. Under the law, the state government is expected to spell out the different areas that such a report ought to cover; for e.g., the quality of water, the extent of water supply in different localities, the measures being taken to prevent contamination, etc.

He said that state governments should set up a regulatory commission, which will have two principal functions. First, the commission should fix rates based on those recommended by the various municipal corporations. Second, it should look into the different costs, assess whether services are being efficiently rendered, and consider plans for improving the quality of services provided. The regulatory commission must also make transparent how much subsidy is being provided to which groups, and whether the state government or the corporation ought to make up the gap between revenue and expenditure. Similarly, the commission must be vested with the authority to monitor from time to time the quality of services being rendered and make recommendations as to the ways in which improvements can be brought about. It should ascertain whether the services provided conform to the state and national standards that are being laid downs; if not, what steps are being contemplated to ensure that these standards are observed. So, on the one hand, the regulatory commission will lead to transparency by making available more information to the people on the range of costs involved and the beneficiaries of various subsidies. On the other, it will act as an independent umpire, which periodically goes into these questions keeping in mind the interests of citizens. Such a commission will provide a mechanism whereby people can give feedback to and interact with the authorities, thus actively shaping the future course of action.

In this context, **Mr. Sivaramakrishnan** pointed out that two organizations – YUVA in Bombay and CIVIC in Bangalore are actively getting involved in the process of formulating municipal budgets. These are among the most mysterious documents in the country, very difficult to read. How does one demystify the municipal budget? How can it be disaggregated so that it is understood at the ward level? To what extent does it live up to the municipality's stated objectives of providing basic services? These were some the questions he raised.

Prof. Atiur Rahman referred to the urban management scenario in Bangladesh. He pointed out that one of the major deficiencies of his country's planning system is that it is unable to integrate 64 districts – which is far less in number than those in India – within the planning process, i.e., preparation, implementation and monitoring of the plan. Without such integration, it is difficult to introduce substantial structural changes in the field of planned development and thereby give a boost to poverty alleviation. Along with urban poverty, he drew attention to rural poverty, observing that the two are strongly interlinked. Citing the example of Dhaka, he pointed out that much of the mismanagement of the city's resources was due to the lack of transparency and monitoring. The political establishment need to play a constructive role in the entire gamut of planning activities – documentation, plan preparation and implementation. Without their positive leadership, not much can be achieved.

Prof. Nazrul Islam observed that major environmental problems basically derive from the mushrooming of private transport. He cited the example of Singapore, where urban pollution and traffic congestion were as bad as today's Dhaka or Delhi. He pointed that Singapore succeeded in emerging from the crisis because of subsidization of and greater investment in public transport and the cap on private transport. He urged governments to subsidize all forms of public transport – waterways, railways, underground – in a transparent manner.

Prof. Islam criticized municipal budgets for being bureaucratic and ad hoc. He shared his experience of a civic movement in Dhaka, which he has been leading for the last five years. His agenda is to demystify budgets by making transparent the mechanisms of the budget-making process. Every year before the national budget, civil society meetings are called. Participatory resource findings on budget making and the prioritization process are presented to all the members. Even the finance minister is invited to address these meetings. This makes the entire process more transparent, accountable and participatory.

Mr. Vijay Saluja emphasized the need to train political personnel in the areas of planning and budget formulation. He pointed out that inefficiency among professionals is not because they do not have the capability; rather their abilities are not put to optimal use. Arguing in favour of tax reforms, he noted that property tax collection is a thankless job if one is not corrupt. An incentive-based approach for motivating personnel engaged in collection of property taxes should be introduced. Arguing against subsidization, he cited the example of New Delhi Municipal Council, which has successfully mobilized resources from land and commercial properties to become financially self-sufficient.

Ms. Alison Barrett noted that while several participants were in favour of involving the community at the micro level, how many examples can be cited where citizens have been involved on a wider scale in big decisions at the city level – whether it is metro railways of major drainage projects. She expressed concern that the institutional process deters public interest litigation.

Mr. Sivaramakrishnan responded to the issue of scaling up. A range of alternatives are available for community-based participation, but it has almost been impossible to scale it up. At the same time, communities cannot always participate. He gave the example of an old lady in slums of Manila. When asked what was her source of livelihood, she said participating in socio-economic surveys. So it is not as if the poor, who too have to earn a livelihood, can easily spare the time for active community participation, in manner that is acceptable to professionals. There is only so much participation that the poor can afford. Further, the participatory mechanism that has been devised, sometimes tend to be an apology for government programmes. How can it be elevated to the city level? Mr Sivaramakrishnan places his trust in the participatory mechanism through ward committees as envisaged in 74t^{h} CAA. Nevertheless, according to him, there are conflicting positions with regard to the ward committee. In West Bengal, there is ward committee for each ward; in Bangalore, there is a ward committee for a group of wards. So there is no consistency. He pointed out that instead of moving to a situation where one can talk about cities without slums, as the UNCHS would like to advocate, the emerging scenario is one of slums without cities. Slum dwellers will soon constitute 30 per cent of the total population. They are already beginning to consolidate their position and political strength – a prominent example are the shack-dwellers of Mumbai – and renegotiating their terms with the political establishment. Today's reality is that the latter is as much dependent on the slum population, which is not a negative feature. It is a reordering of negotiating terms between the leadership and the public. This, in a way, is a human rights approach to city ordering. Most of our cities today and, those of tomorrow, are/will be unintended cities. The unintended city does not respond to intended laws and procedures but devises its own set.

On the question of a pluralist political structure, Mr Sivaramakrishnan noted that across the world, there are not many examples of pluralist societies. There is a tendency towards homogeneity, enforced by the market. In such a context, how can the political leadership be trained? In a broadly elective system, with a very large number of representatives, how does society find a way to intelligently and cogently engage them? In his views, at least a limited section of the new tier of leadership would have genuine concerns. Even if a small percentage, say 30 per cent of the 60,000 elected urban representatives in India, ask the questions why are we here, what are we supposed to do, sooner or later, the chemistry of interaction is bound to change. This process is already unfolding, though the pace is slow. So it is important to keep on engaging people. Mr. Sivaramakrishnan expressed happiness in the fact that at least in West Bengal, one got the impression that professionals and the political leadership do not necessarily pussy-foot around. There seems to be more open dialogue and exchange of ideas in this state compared to other states, which is extremely positive. As he put it, if they listen to you today, they may act upon it tomorrow. Similarly, if you listen to them today, you may revise your views of yesterday.

Prof. Golam Rahman said that in Dhaka, the town plan is prepared by the capital development authority or RAJUK. The mayor doesn't have any role at all. Previously, when RAJUK was called Dhaka Improvement Trust (DIT), there were elected commissioners on the committee. Now, RAJUK is only a six-member committee. All the members are government appointees, and none of them has training in the field of planning and development. Earlier, judges of the Supreme Court, inspectors general of police, representatives of civil society, professionals from trade and commerce served on the committee. So RAJUK used to be a well-represented body. Now it is exclusively a government nominated body, which is under the Ministry of Works and Housing. The mayor is accountable to the Ministry of Local Government, Rural Development and Cooperatives. So two separate ministries are involved. They work in a parallel fashion, rarely coordinating with each other. In such a top-down approach, is there any scope for bringing local-level politicians into the planning process? But they are the ones who are supposed to know their area very thoroughly – where does the water clog, which drainage is faulty, what are the problems of the slum dwellers, how serious are the fire hazards, etc. Since they go to each and every household during elections for votes, it is important to involve them in the task of planning through training programmes.

Prof. Nazrul Islam explained the various models of planning that exist in Bangladesh. He too criticized RAJUK for being a techno-bureaucratic organization manned by government officials. However, the reverse work culture prevails in the water and sewerage authority of Dhaka. Here the board consists of 12 members, 10 of whom represent civil society and professional groups. Only two government ministries find representation; consumers are also represented. This new form of governance took shape on the insistence of the World Bank. The Bank had provided financial assistance of 7500 million taka for a water treatment plant. The funds were made available on the condition that a more democratic system of governance would be introduced. This came about by an act of Parliament. So there are different models of participation. The Healthy City Programme of World Health Organization, CDS (City Development Strategy) by the World Bank and the UNICEF models for the delivery of urban basic services are examples of active participation at the grassroots level and of strong leadership given by the elected ward commissioners. Some of them come on their own; others have to be inducted into the system. These initiatives need to be linked up with city-level efforts. In case of Dhaka, the mayor of Dhaka City Corporation does participate. In some cases, he comes spontaneously, with enthusiasm; in other instances, he has to be brought in! The case of Chittagong is unique. The Mayor of Chittagong spares no effort – he even tries to negotiate with foreign governments – for improving the city. Thus, integration of actions and measures at different levels is possible. This is happening in Bangladesh in the case of urban basic services.

Mr. Tapan Nath also talked about people's participation in Dhaka city. He pointed out that citizens are coming forward as a countervailing force. Not only are they protesting against the dismal performance of the city government, they are also initiating positive actions. In 1999, there was a sudden outbreak of dengue fever in Dhaka. Hundreds of people suffered, many succumbed to the disease. At that time, people came forward to create awareness among their fellow citizens on how to combat dengue. They worked very closely with the city corporation. Through voluntary action, they cleared many parts of the city that had become breeding grounds for mosquitoes. Schools were also involved in cleaning up the city; and this has virtually been institutionalized. The private sector contributed funds for the awareness campaign against dengue. This resulted in a much lower incidence of dengue in 2001.

He also mentioned the 'Save Buriganga Movement'-- a campaign by the people to save the Buriganga river from pollution and encroachment. Another people's initiative is on air pollution. Vehicles with the killer two-stroke engines are severely polluting Dhaka, with children the most affected. This health hazard has become one of the leading problems of the city. Civil society groups, including NGOs, are raising their voice through rallies and other measures. With public pressure mounting, the government is thinking of getting all vehicles in the city to switch to CNG.

Dr. Atiur Rahaman spoke on the planning experience of the city of Chittagong in Bangladesh. He expressed reservations about technology transfer. While most often high technology is used for documentation purposes and to collect data, it does not live up to its promise during the implementation stage. This, he argued, requires political commitment and a holistic approach to planning. The healthy cities movement has been approached from this perspective and can contribute to the involvement of the community and other development partners in a city.

Dr. Ranjit Perera talked about the success of the Colombo Municipal Council (CMC), a presentation which had been made by Mr Omar Kamil, the Mayor of (CMC). He identified political willingness as the key factor in the success of the programme. The commitment of the mayor, both the previous one and the present incumbent, to change the style of governance was a decisive factor. Being a very prominent figure in the private sector, the previous mayor brought a different management style to the municipal council. One of the very first things that he did was to invite representatives from different interest groups for a meeting, which included citizens, professionals, academics, leaders in the private sector, those heading social organizations, like Lions Club or Rotary etc. Discussions were held on some of the most pressing issues faced by the city. These initiatives were taken much before the CDS programme, or the urban management programme of the UNDP or UNCHS started in Colombo. One of the proposals was to invite the private sector to participate in the development of the city. It was involved in the planning of streets, regularization of their names and other simple things, like putting up signboard, etc. These initiatives had a high visual impact – something which the private sector was obviously very keen on. Similarly, it got actively involved in garbage management. Working with the private sector, in turn, influenced local government officers. For long, given to doing routine work, they needed to become more proactive in the new climate of competition. Dr. Perera concluded by listing the key factors that contributed to Colombo Municipal Corporation's success: (*a*) political willingness; (*b*) stakeholders' active participation; (*c*) visibility of the change; and (*d*) motivation of municipal officers to do things in a different manner.

Abbreviations

AMU	-	Appraisal and Monitoring Unit
AP	-	Andhra Pradesh
APUSP	-	Andhra Pradesh Urban Services for the Poor
BAPA	-	Bangladesh Paribesh Andolon
BCSIR	-	Bangladesh Council of Scientific and Industrial Research
BELA	-	Bangladesh Environmental Lawyers Association
BMA	-	Bangkok Metropolitan Administration
BMRDA	-	Bombay Metropolitan Regional Development Authority
BOLT	-	Build Own Lease Transfer
BOO	-	Build Operate Own
BOOT	-	Build Own Operate Transfer
BOT	-	Build Operate Transfer
BPL	-	Below Poverty Level
BOD	-	Biochemical Oxygen Demand
BUET	-	Bangladesh University of Engineering Technology
CAA	-	Constitutional Amendment Act
CAG	-	Comptroller and Auditor General
CAP	-	Community Action Planning
CB-GIS	-	Community Based Geographical Information System
CBO	-	Community Based Organisation
CDC	-	Community Development Committee
CDC	-	Centre for Development Communication
CDC	-	Community Development Council
CDS	-	Community Development Society
CDS	-	City Development Strategies
CEE	-	Centre for Environment Education
CETP	-	Common Effluent Treatment Plant
CIDCO	-	City and Industrial Development Corporation
CII	-	Confederation of Indian Industry
CMC	-	Colombo Municipal Council
CMIE	-	Centre for Monitoring Indian Economy
CNG	-	Compressed Natural Gas
COD	-	Chemical Oxygen Demand
COVA	-	Conference of Voluntary Agencies
CPCB	-	Central Pollution Control Board
CSE	-	Centre for Science and Environment

CSIP	-	Calcutta Slum Improvement Project
CSIR	-	Council for Scientific and Industrial Research
CSO	-	Civil Society Organisation
CUS	-	Centre for Urban Studies
DCC	-	Dhaka City Corporation
DFID	-	Department for International Development
DLB	-	Directorate of Local bodies
DMDP	-	Dhaka Metropolitan Development Plan
DPC	-	District Planning Committee
DSK	-	Dustha Sastha Kendra
DSMA	-	Dhaka Statistical Metropolitan Area
DWASA	-	Dhaka Water and Sewerage Authority
DWCUA	-	Development of Women and Children in Urban Areas
E & TL	-	Environmental Traffic Light
EPA	-	Environment Protection Act
EPM	-	Environmental Management Programme
FICCI	-	Federation of Indian Chamber of Commerce and Industries
FIRE	-	Financial Institutional Reforms and Expansion
GoB	-	Government of Bangladesh
GoI	-	Government of India
HMWS & SB	-	Hyderabad Metropolitan Water Supply and Sewerage Board (HMWS & SB)
HUDA	-	Hyderabad Urban Development Authority
HUDCO	-	Housing and Urban Development Corporations
ICBEN	-	International Conference on Bangladesh Environment
ICLEI	-	International Council for Local Environment Initiative
IES	-	Indian Environment Society
IHS	-	Institute of Housing Studies
IIT	-	Indian Institute of Technology
IL&FS	-	Infrastructure Leasing & Financial Services
ILGUS	-	Institute of Local Government and Urban Studies
IMF	-	International Monetary Fund
IMR	-	Infant Mortality Rate
IPP	-	India Population Project
ISI	-	Indian Standards Institute
KEIP	-	Kolkata Environment Improvement Project
KIP	-	Kampung Improvement Programme
KMA	-	Kolkata Metropolitan Area

KMC	-	Kolkata Municipal Corporation
KMDA	-	Kolkata Metropolitan Development Authority
KUSP	-	Kolkata Urban Services for the Poor
LA 21	-	Local Agenda 21
LEAP	-	Local Environmental Action Plan
LGED	-	Local Government Engineering Department
LGRDC	-	Ministry of Local Government, Rural Development and Cooperatives
LIC	-	Life Insurance Corporation
LIFE	-	Local Initiatives for Environment
LNG	-	Liquid Natural Gas
LPI	-	Local Partner Institute
MAPP	-	Municipal Action Plan for Poverty Reduction
MCH	-	Municipal Corporation of Hyderabad
MCL	-	Municipal Corporation of Ludhiana
MDP	-	Municipal Development Programme
MED	-	Municipal Engineering Directorate
MFC	-	Municipal Finance Commission
MHPW	-	Ministry of Housing and Public Work
MIC	-	Mayour – in – Council
MMRDA	-	Mumbai Metropolitan Regional Development Authority
MOU	-	Memorandum of Understanding
MPC	-	Metropolitan Planning Committee
MRC	-	Municipal Reforms Committee
MSU	-	Municipal Strengthening Unit
MTF	-	Municipal Task Force
NCU	-	National Commission on Urbanisations
NEERI	-	National Environmental Engineering Research Institute
NESDB	-	National Economic and Social Development Board
NGO	-	Non-government Organisation
NH	-	National Highway
NHG	-	Neighbourhood Group
NIPFP	-	National Institute of Public Finance and Policy
NIUA	-	National Institute of Urban Affairs
NSDP	-	National Slum Development Programme
OECD	-	Organisation of Economic Cooperation and Development
O & M	-	Operation and Maintenance
OPD	-	Out Patients Department
OPP	-	Orangi Pilot Project

PEEP	-	Primary Education Enhancement Project
PHED	-	Public Health Engineering Department
PO	-	Private Organisations
POROSH	-	Paribesh Rokkha Sopath
PPA	-	Participatory Poverty Assessment
PSP	-	Private Sector Participation
RAJUK	-	Capital Development Authority
RCV	-	Resident Community Volunteers
RUEA	-	Rapid Urban Environmental Assessment
SC	-	Scheduled Castes
SCP	-	Stakeholders Consultative Process
SCP	-	Sustainable Cities Programme
SDO	-	Structured Debt Obligation
SEMP	-	Sustainable Environmental Management Programme
SFC	-	State Finance Commission
SINPA	-	Sustainable Implementation of National Plane of Action
SJSRY	-	Swarna Jayanti Shahari Rojgar Yojona
SKAA	-	Sindh Katchi Abadi Authority
SMKML	-	Sangu Miraj Kupwad Municipal Council
SP	-	Strategic Planning
SPA	-	Special Planning Authority
SPAPC	-	Society for the Promotion of Area Resource Centres
SSEP	-	Small Scale Enterprises Programme
ST	-	Scheduled Tribes
SWASTI	-	Solid Waste Initiatives
SWM	-	Solid Waste Management
SWM	-	Solid Waste Management
SWOT	-	Strengths, Weaknesses, Opportunities and Threats
TERI	-	Tata Energy Research Institute
TWINS	-	Twin Cities Integrated Network Services
UBSP	-	Urban Basic Services for the Poor
UEM	-	Urban Environment Management
UIF	-	Urban Initiative Fund
ULB	-	Urban Local Bodies
ULCRA	-	Urban Land (Ceiling and Regulations) Act
UMP	-	Urban Management Programme
UNCED	-	United Nations Conference on Environment and Development
UNCHS	-	United Nations Centre for Human Settlements

UNDP	-	United Nations Development Programme
UNEP	-	United Nation Environment Programme
UNIRSD	-	United Nations Institute for Research and Social Development
USAID	-	United Nations Agency for International Development
US-AEP	-	United States-Asia Environmental Partnership
USEP	-	Urban Self-Employment Programme
UWEP	-	Urban Wage Employment Programme
VVSR	-	Vasai – Virar Sub-region
WHO	-	World Health Organisation
WPC	-	Western Provincial Council
WTO	-	World Trade Organisation

INDEX